Explorations in Time-Frequency Analysis

An authoritative exposition of the methods at the heart of modern nonstationary signal processing from a recognized leader in the field. Offering a global view that favors interpretations and historical perspectives, it explores the basic concepts of time-frequency analysis and examines the most recent results and developments in the field in the context of existing, lesser-known approaches. Several example waveform families from bioacoustics, mathematics, and physics are examined in detail, with the methods for their analysis explained using a wealth of illustrative examples. Methods are discussed in terms of analysis, geometry, and statistics. This is an excellent resource for anyone wanting to understand the "why" and "how" of important methodological developments in time-frequency analysis, including academics and graduate students in signal processing and applied mathematics, as well as application-oriented scientists.

Patrick Flandrin is a research director at the CNRS (Centre national de la recherche scientifique), working in the Laboratoire de Physique of the École normale supérieure de Lyon. He is a Fellow of the IEEE and EURASIP, and a Member of the French Academy of Sciences.

Explorations in Time-Frequency Analysis

PATRICK FLANDRIN

École normale supérieure de Lyon

CAMBRIDGE
UNIVERSITY PRESS

University Printing House, Cambridge CB2 8BS, United Kingdom

One Liberty Plaza, 20th Floor, New York, NY 10006, USA

477 Williamstown Road, Port Melbourne, VIC 3207, Australia

314–321, 3rd Floor, Plot 3, Splendor Forum, Jasola District Centre, New Delhi – 110025, India

79 Anson Road, #06–04/06, Singapore 079906

Cambridge University Press is part of the University of Cambridge.

It furthers the University's mission by disseminating knowledge in the pursuit of
education, learning, and research at the highest international levels of excellence.

www.cambridge.org
Information on this title: www.cambridge.org/9781108421027
DOI: 10.1017/9781108363181

© Cambridge University Press 2018

First published 2018

Printed in the United Kingdom by TJ International Ltd. Padstow Cornwall

A catalogue record for this publication is available from the British Library.

Library of Congress Cataloging-in-Publication Data
Names: Flandrin, Patrick, author.
Title: Explorations in time-frequency analysis / Patrick Flandrin
    (Ecole normale superieure de Lyon).
Description: Cambridge, United Kingdom ; New York, NY : Cambridge University Press, 2018. |
    Includes bibliographical references and index.
Identifiers: LCCN 2018010021 | ISBN 9781108421027 (hardback) |
    ISBN 1108421024 (hardback)
Subjects: LCSH: Signal processing–Mathematics. | Time-series analysis. | Frequency spectra.
Classification: LCC TK5102.9 .F545 2018 | DDC 621.382/20151955–dc23
    LC record available at https://lccn.loc.gov/2018010021

ISBN 978-1-108-42102-7 Hardback

To Marie-Hélène, Lou, Margot, and Zoé,
for yesterday, for today, and for tomorrow.

Contents

Acknowledgments   *page* x
Preface   xiii
Notation   xv

1   Introduction   1

Part I Basics and Constraints   7

2   Small Data Are Beautiful   9
   2.1   Gravitational Waves   9
   2.2   Bats   11
   2.3   Riemann-Like Special Functions   14
   2.4   Chirps (Everywhere)   16

3   Of Signals and Noise   21
   3.1   Order versus Disorder   21
   3.2   Signals   22
   3.3   Noise   24

4   On Time, Frequency, and Gauss   29
   4.1   Gauss   29
   4.2   From Gauss to Fourier   31
   4.3   From Gauss to Shannon-Nyquist   31
   4.4   From Gauss to Gabor   32

5   Uncertainty   35
   5.1   Variance   35
   5.2   Entropy   38
   5.3   Ubiquity and Interpretation   39

6   From Time and Frequency to Time-Frequency   40
   6.1   Correlation and Ambiguity   40
   6.2   Distribution and Wigner   43
   6.3   Spectrograms, Cohen, and the Like   46

7   **Uncertainty Revisited**   50
    7.1 L_2-Norm   50
    7.2 L_p-Norms and Entropy   51
    7.3   Concentration and Support   51
    7.4   Variance   53
    7.5   Uncertainty and Time-Frequency Localization   54

8   **On Stationarity**   56
    8.1   Relative Stationarity   57
    8.2   Testing Stationarity   60

Part II Geometry and Statistics   67

9   **Spectrogram Geometry 1**   69
    9.1   One Logon   69
    9.2   Two Logons   70
    9.3   Many Logons and Voronoi   73

10   **Sharpening Spectrograms**   77
    10.1 Reassignment   78
    10.2 Multitaper Reassignment   83
    10.3 Synchrosqueezing   88
    10.4 Sparsity   90
    10.5 Wedding Sharpening and Reconstruction   96

11   **A Digression on the Hilbert–Huang Transform**   98
    11.1 Empirical Mode Decomposition   98
    11.2 Huang's Algorithm   100
    11.3 The Hilbert–Huang Transform   100
    11.4 Pros, Cons, and Variations   101

12   **Spectrogram Geometry 2**   106
    12.1 Spectrogram, STFT, and Bargmann   106
    12.2 Reassignment Variations   107
    12.3 Attractors, Basins, Repellers, and Contours   111

13   **The Noise Case**   116
    13.1 Time-Frequency Patches   116
    13.2 Correlation Structure   118
    13.3 Logon Packing   121

14   **More on Maxima**   124
        14.1 A Randomized Lattice Model   124
        14.2 Ordinates and Maxima Distributions   129
        14.3 Voronoi   134

15   **More on Zeros**   139
        15.1 Factorizations   139
        15.2 Density   143
        15.3 Pair Correlation Function   144
        15.4 Voronoi   145
        15.5 Delaunay   149
        15.6 Signal Extraction from "Silent" Points   153
        15.7 Universality   161
        15.8 Singularities and Phase Dislocations   164

16   **Back to Examples**   168
        16.1 Gravitational Waves   168
        16.2 Bats   175
        16.3 Riemann-Like Special Functions   188

17   **Conclusion**   197

18   **Annex: Software Tools**   199

        References   201
        Index   210

Acknowledgments

The time-frequency explorations reported in this book have been shared for more than 30 years with a number of people that I am particularly pleased to thank for their collaboration, their criticisms (including comments on earlier drafts of this book, that I did my best to incorporate), and their friendship. Those include Paulo Gonçalves, Patrice Abry, Olivier Michel, Richard Baraniuk, and Franz Hlawatsch, with whom I had many time-frequency interactions in the early days of the SISYPH (Signals, Systems, and Physics) group in Lyon, and still have always fruitful discussions even if our scientific paths may have somehow moved apart; Éric Chassande-Mottin and François Auger, with whom we tirelessly pursued the quest for reassignment and its avatars, taking advantage of point-wise collaborations with Ingrid Daubechies and, more recently, Hau-tieng Wu; Pierre Borgnat and Nelly Pustelnik, who, among many other things, made compressed sensing and optimization enter the field; Cédric Richard, Paul Honeine, Pierre-Olivier Amblard, and Jun Xiao, with whom we revisited stationarity within the StaRAC project; Sylvain Meignen, Thomas Oberlin, Dominique Fourer, Jinane Harmouche, Jérémy Schmitt, Stephen McLaughlin, and Philippe Depalle, who helped opening new windows on signal disentanglement within the ASTRES project; Gabriel Rilling, Marcelo Colominas, Gaston Schlotthauer, and Maria-Eugenia Torres, who joined me (and Paulo Gonçalves) in the exploration of the foggy land of Empirical Mode Decomposition; Pierre Chainais and Rémi Bardenet, with whom I started discovering the fascinating territories of determinantal point processes and Gaussian Analytic Functions; and finally Géraldine Davis for her fresh and sharp eye.

A special mention is to be made of Yves Meyer: I had the privilege of meeting him during the earliest days of the wavelet adventure and benefiting from his scientific and human stature – not to mention his faultless support. I also thank Odile Macchi, whose work has (almost) nothing to do with the content of this book, but whose personality and scientific trajectory have always been an example for me, and whose day-to-day friendship is invaluable.

As a full-time researcher of CNRS (Centre national de la recherche scientifique, i.e., the French National Council for Scientific Research), I had the opportunity to pursue a research program in total freedom and the possibility of envisioning it in the longrun. I appreciate this privilege, as well as the excellent conditions I always found at École normale supérieure de Lyon and its Physics Department: it is my pleasure to warmly thank both institutions. Some recent aspects of the work reported here have been developed within the StaRAC and ASTRES projects, funded by Agence nationale de la recherche (under respective grants ANR-07-BLAN-0191 and ANR-13-BS03-0002-01), whose support is gratefully acknowledged. Finally, I would like to thank the Isaac Newton Institute for Mathematical Sciences, Cambridge, for support and hospitality during the "Statistical Network Analysis" program (supported by EPSRC grant no EP/K032208/1) where a large portion of the writing of this book has been undertaken and accomplished.

When not far from Internet and e-mail distractions in quiet Montgrenier, the rest of the writing was completed in Lyon, and in particular at a back table of Café Jutard, which is also gratefully acknowledged for its atmosphere and coffee quality.

Preface

Paterson: [voice over]
"I go through
trillions of molecules
that move aside
to make way for me
while on both sides
trillions more
stay where they are."
　　　　　　—Ron Padgett

Time-frequency can be considered the natural language of signal processing. We live in an ever-changing world, with time as a marker of events and evolutions. And from light to gravitation, from biological clocks to human activities, our everyday experience is faced with waves, oscillations, and rhythms, i.e., with frequencies.

Thinking of audio, the first attempts to record and transcribe the human voice go back to 1857, with the invention of the "phonautograph" by Édouard-Léon Scott de Martinville [1, 2]. The recording of speech as a signal was a success, but the transcription from the only waveform was a failure. The modern way out only came almost one century later with the invention of the "sound spectrograph" by W. Koenig, H. K. Dunn, and D. Y. Lacy [3], who in 1946 opened a new window on speech analysis by unfolding waveforms into time-frequency images. Starting therefore with "visible speech" [4] questions, the applications of time-frequency analysis later happened to be not only unlimited, but also instrumental in almost all areas of science and technology, culminating somehow in the pivotal role recently played by wavelet-like methods in the first detection of gravitational waves [5, 6].

The development of time-frequency analysis that has been observed since the 1980s has led to the writing of many articles as well as of a number of books, including by the present author [7], and one may question the need to add one more piece to the existing literature. In fact, some of the books written at the turn of the century were originally research monographs, and over the years, their content has either become standard material or has been sidestepped or superseded by new developments. New achievements have appeared and, for newcomers to the field, have been adopted mainly in comparison to the current state of the art, which itself resulted from a cumulative construction that most often lost track of the earliest attempts. One of the motivations for the writing of the present book is therefore to be found in the desire to offer the reader an overview of key concepts and results in the field by bridging new advances and older ideas that, even if they have not been fully followed per se, have been instrumental in deriving the techniques that were eventually adopted. In doing so, this book presents a series of *explorations* that mix elementary, well-established facts that are at the core of time-frequency analysis with more recent variations, whose novelty can be rooted in much older ideas.

Notation

STFT	Short-Time Fourier Transform
EMD	Empirical Mode Decomposition
AM	Amplitude Modulation
FM	Frequency Modulation
1D, 2D, 3D	one-dimensional, two-dimensional, three-dimensiional
$\mathbb{R}$	real numbers
$\mathbb{C}$	complex numbers
i	$\sqrt{-1}$
Re$\{z\}$	real part of $z \in \mathbb{C}$
Im$\{z\}$	imaginary part of $z \in \mathbb{C}$
$\mathbb{P}(.)$	probability function
$p(.)$	probability density function
$\mathbb{E}\{X\}$	expectation value of X
var$\{X\}$	variance of X
cov$\{X, Y\}$	covariance of X and Y
$\mathcal{H}(p)$	Shannon entropy of probability density function p
$\mathcal{H}_\alpha(p)$	α-order Rényi entropy of probability density function p
t	time
ω	(angular) frequency
$x(t)$, $X(\omega)$	Fourier transform pair
$x^*(t)$	complex conjugate of $x(t)$
E_x	energy of $x(t)$
τ	time delay
ξ	Doppler shift
$\|x\|_p$	L_p-norm of $x(t)$
$L^p(\mathbb{R})$	space of L_p-integrable functions
$\langle x, y \rangle$	inner product of $x(t)$ and $y(t)$
$\overline{F}^\rho$	average of F with respect to the density ρ
$\delta(.)$	Dirac's "δ-function"
δ_{nm}	Kronecker's symbol (= 1 if $n = m$ and 0 otherwise)
$e_\omega(t)$	monochromatic wave of frequency ω
$\mathbf{1}_T(t)$	indicator function of interval T (= 1 if $-T/2 \leq t \leq +T/2$ and 0 otherwise)
$\gamma_x(\tau)$	stochastic correlation function of $x(t)$
$\tilde{\gamma}_x(\tau)$	deterministic correlation function of $x(t)$

$\Gamma_x(\omega)$	spectrum density of $x(t)$
$r_x(\tau)$	relation function of $x(t)$
$(\mathbf{F}x)(t)$	Fourier transform of $x(t)$
$(\mathbf{H}x)(t)$	Hilbert transform of $x(t)$
$(\mathbf{M}x)(t)$	Mellin transform of $x(t)$
$(\mathbf{T}_{\tau,\xi}x)(t)$	time-frequency shift operator acting on $x(t)$
$h_k(t)$	Hermite function
$g(t)$	circular Gaussian window
$\mathcal{F}_x(z)$	Bargmann transform of $x(t)$
$F_x^{(h)}(t,\omega)$	STFT of $x(t)$ with window $h(t)$
$S_x^{(h)}(t,\omega)$	spectrogram of $x(t)$ with window $h(t)$
$\hat{S}_x^{(h)}(t,\omega)$	reassigned spectrogram of $x(t)$ with window $h(t)$
$\hat{t}(t,\omega), \hat{\omega}(t,\omega)$	reassignment time, reassignment frequency
$\hat{\mathbf{r}}_x(t,\omega)$	reassignment vector field of $x(t)$
$\tilde{F}_x^{(h)}(t,\omega)$	synchrosqueezed STFT of $x(t)$ with window $h(t)$
$W_x(t,\omega)$	Wigner distribution of $x(t)$
$A_x(\xi,\tau)$	ambiguity function of $x(t)$
$C_x(t,\omega;\varphi)$	Cohen's class distribution of $x(t)$ with kernel $\varphi(\xi,\tau)$

1 Introduction

One buzzword that has gained in popularity since the beginning of this century is "data science." What data science actually is, however, is a matter of debate (see, e.g., [8]). It can be argued that, since its objectives are to collect, analyze, and extract information from data, data science is ultimately a revamping of "statistics." A case could also be made for adding "signal processing" to the picture since, according to IEEE (the flagship society for the field), signal processing is "the science behind our digital life." As such, the very broad scientific coverage of signal processing makes it difficult to delineate for it a clear borderline with data science, which itself appears as one of the items listed by IEEE.

Whatever the terminology, the "data science/signal processing" main issue can be summarized as follows:

> Starting from some *data* (be they human-made or given by nature), the objective is to extract from them some information, assumed to exist and considered of interest.

Remark. The question of what is "of interest" or not makes the end user enter the process. A common distinction is made in signal processing between "signal" and "noise" (we will come back to this in Chapter 3), but it must be understood that this has only a relative meaning. For instance, in the case of the famous *cocktail-party problem*, which can be formulated in terms of source separation, it is clear that, when considering one specific conversation, this one becomes a "signal" while other conversations, although they are of the very same nature, are "noise." In a nutshell, signals to some are noise to others. Another example is given by the so-called *Cosmic Microwave Background* that can be seen as either a perturbation for communications (this was even the way it was discovered by Arno Penzias and Robert W. Wilson in 1965) or a scientific object per se, the analysis of which gives invaluable information about the early universe [9].

The process of extracting information from data encompasses many facets that may include acquisition, transformation, visualization, modeling, estimation, or classification. Signal processing (or data analysis) has therefore something to do with three main domains, namely *physics*, *mathematics*, and *informatics*. First, physics, which has to be understood at large, i.e., as in direct connection with the physical world where data live and/or are originated from (from this point of view, this also includes biological or even symbolic data). Then, mathematics (including statistics), which permits us to formalize transforms and manipulations of data, as well as assessing the performance of analysis and processing methods. And finally, informatics, which helps dealing with digitized data and turns processing methods into algorithms.

Remark. Signal processing is interdisciplinary by nature, making more difficult to appreciate its specificity when referring to classical categorizations of science (in the spirit of, say, Auguste Comte). Indeed, it took a while for it to be recognized as a discipline of its own. In its early days, i.e., during World War II and right after, signal processing was not necessarily named as such and was mostly considered through its applied and/or technological aspects, despite theoretical breakthroughs such as Norbert Wiener's "yellow peril" seminal book, which first took the form of a classified report in 1942 and was eventually published as a book in 1949 [10]. Though still facing practical problems (raised in particular by the U.S. Navy and the French Navy about background noise in underwater acoustics), several efforts were then pursued for giving signal processing solid grounds by bridging physics and mathematics. One landmark book in this respect is *Théorie des Fonctions Aléatoires* (i.e., *Theory of Random Functions*) by André Blanc-Lapierre and Robert Fortet [11]. This pioneering book established the field and launched a successful French school that later, in 1967, organized the first ever Groupe d'Etudes du Traitement du Signal (GRETSI) conference – the first one of the series of GRETSI symposia that are still run every two years – that was specifically dedicated to signal processing. This was followed about 10 years later by IEEE, which held its first International Conference on Acoustics, Speech and Signal Processing, or ICASSP, conference in 1976. Roughly speaking, considering countless conferences, books, and periodicals that are now flourishing, one can say that signal processing emerged as a recognized field during the 1970s.

In short, signal processing exists at the intersection of different fields, reducing to none of them but gaining from their confrontation an intrinsic value that goes beyond a simple addition of each. In "complex systems" terminology, we would say that "the whole is more than the sum of the parts" and, as illustrated in Figure 1.1:

> We claim that the success of a signal processing/data analysis method, as well as its acceptance by the scientific community, is based on a well-balanced importance of the three components of the "golden triangle" whose vertices are physics (data), mathematics (formalizations and proofs), and informatics (algorithms).

Remark. A companion interpretation of a similar "golden triangle" can be given by attaching to the bottom-right vertex the possibilities offered by informatics in terms of *simulation*. In such a case, the balance is among classical experiments rooted in physics, models expressed in mathematical terms, and numerical data generated by model-based equations governed by physics.

Let us support our claim about the "golden triangle" of Figure 1.1 by considering two examples that are closely related to the purpose of this book and to the methods that will be considered (we will later discuss some counterexamples).

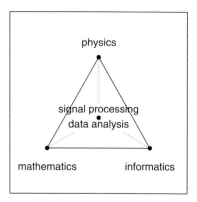

Figure 1.1 The "golden triangle." This symbolic representation states that signal processing (or data analysis) is based on interactions among three basic components rooted in the more classical fields of physics (data), mathematics (formalizations and proofs), and informatics (algorithms).

Example 1 – Fourier. Let us start with the first success story, namely Fourier analysis, which is of considerable importance in signal processing. The starting point is clearly physics, since the purpose of Joseph Fourier's seminal work (completed in 1811, yet published only in 1822 [12]) was to answer the question, raised by the French Academy of Sciences, of establishing an analytic theory of heat. To this end, Fourier developed a mathematical technique of expansions based on sines and cosines that is commonly used today (Fourier series and integrals), and launched the whole field of harmonic analysis, which experienced tremendous developments during the nineteenth and twentieth centuries. Because of its potential in many applications, it soon became natural to look for "implementations" of Fourier analysis, leading to the invention of mechanical, optical, acoustical, and electrical devices [13]. Following the electronic revolution and the advent of computers, however, the true breakthrough came from algorithmics with the publication of the *Fast Fourier Transform* by James Cooley and John Tukey in 1965 [14]. Besides physics (with an unlimited range of applications to all kinds of data) and mathematics, this was the third key ingredient that really boosted Fourier analysis and made it enter the toolkit of every scientist. Incidentally, it is worth remarking that this point of view, which considers the three mentioned aspects altogether, was already at the heart of Fourier's program, as attested by two quotes [15] excerpted[1] from his *Théorie Analytique de la Chaleur* [12]. The first one is quite famous and states that "The deep study of nature is the most fruitful source of mathematical discoveries." The second one is less known, but nonetheless visionary, since it reads: "This difficult research required a special analysis, based on new theorems [...]. The proposed method ends up with nothing vague and undetermined in its solutions; it drives them to their ultimate numerical applications, a condition which is necessary for any research, and without which we would only obtain useless transformations." Nothing to add: physics, mathematics, numerics – all three are equally necessary in Fourier's words.

---

[1] The translation from French to English is mine.

Example 2 – Wavelets. The second example is quite similar to Fourier analysis, and it is in some sense one of its avatars. As for Fourier and heat theory, *wavelet theory* also originated from a physics problem, namely the question of *vibroseismics* in geophysics. The technique, used for oil exploration, basically amounts to sending mechanical vibrations of increasing frequency into the ground from the surface and analyzing the returning echoes to produce information about the underlying geological structure. Because of the nonstationary nature of the excitation, Jean Morlet, a French engineer working for the Elf-Aquitaine group, was keen to use a time-frequency analysis, but he realized that the standard way, namely the Gabor expansion, had some shortcomings. Some of them were of a computational nature (typically, unstable reconstructions) and other ones came from physical considerations. Morlet's major objection to Gabor analysis was that it is based on a window of fixed duration, whatever the analyzed frequency. This means that, when analyzing "high" frequencies, the window may contain many oscillations, whereas when going down to "low" frequencies, the same window may contain only a fraction of one oscillation, questioning the concept of frequency itself. This echoes a remark of Norbert Wiener in his autobiography *Ex-Prodigy: My Childhood and Youth* [16]: "A fast jig on the lowest register of the organ is in fact not so much bad music as no music at all."

This physical observation prompted Morlet to look for a better mathematical decomposition, and his proposal was to forget about a duration-invariant window with a variable number of oscillations, preferring a *shape-invariant* waveform made of a few oscillations (hence the name "wavelet"), whose duration would be *locked* to the (inverse) analyzed frequency. This seemingly simple idea was first developed with Alex Grossmann from a mathematical physics point of view [17]. Soon after, it became a major topic in mathematics thanks to the pioneering works of Yves Meyer [18], Ingrid Daubechies [19], and Stéphane Mallat [20] (to name but a few), who put theory on firm and elegant grounds. As for Fourier and the Fast Fourier Transform (FFT), the impact of wavelets has been leveraged when wedding mathematics with electrical engineering, recognizing that wavelet bases can be given a filter bank interpretation, and that wavelet transforms can be equipped with fast and efficient algorithms [20]. Again, starting from physics and getting a proper mathematical formalization, closing the "golden triangle" with computational efficiency was instrumental in adopting and spreading wavelets in almost every domain of science and technology.

Coming back to data science in general, a current trend is to think "big." Indeed, we are now overwhelmed by a deluge of data that may take a myriad of forms and dimensionalities, be they multivariate, multimodal, hyperspectral, non-euclidian, or whatever. This has created a move of data analysis from classical signal processing or time series analysis toward new avenues that are paved with buzzwords such as data mining, large-scale optimization, or machine learning (preferably "deep"). Of course, the point is not to question those approaches that led to tremendous success stories. It is rather to consider that there still remains some room for a more "entomological" study of the fine structure of modest size waveforms, calling in turn for some "surgical" exploration of the methods dedicated to their analysis. This is what this book is about.

> In this era of "big data," we propose to think about signals another way, to think "small"!

To achieve this program, a specific perspective will be adopted, namely that of describing signals *jointly in time and frequency*; this is similar to the method of analysis used in the wavelet technique mentioned previously, but is not restricted to only this type of analysis. It is well known that a signal most often takes the form of a "time series," i.e., a succession of values that reflect the temporal evolution of some quantity. This may concern speech, music, heartbeats, earthquakes, or whatever we can imagine as the output of some sensor. Most data often involve rhythms, cycles, and oscillations; it is also well known that there exists a powerful mathematical tool, the *Fourier transform*, that allows for a complementary description of the very same data in a dual, frequency-based domain. This yields a different yet equivalent representation, and we can go back and forth from one to the other without losing any information. However, what each of those representations tells us about some data is not only of a different nature but also *orthogonal* in the sense that they are exclusive of each other: a frequency spectrum just ignores time, mirroring the "natural" representation in time that makes no direct reference to a frequency content.

As powerful as it has proven to be from a mathematical point of view, this alternative contrasts with our everyday experience, which seems to indicate that time and frequency should be able to interact and exchange information; after all, whistling does seem to make frequency vary with time. While this is an idea that nobody really has a problem with, it is something that a strict Fourier representation cannot easily handle. The classic analogy that is used when speaking of overcoming this Fourier limitation is often that of a *musical score*, i.e., a symbolic representation that makes use of two dimensions for writing down a musical piece: time on the one hand for the occurrence and duration of different notes, and frequency on the other hand for their pitch. It should be noted, however, that a musical score is a *prescription* for what a signal (the musical piece when actually played) should be.

> *Time-frequency* analysis goes somehow the other way, its very purpose being the *writing* of the musical score, given a recording.

There is unfortunately no unique and completely satisfactory way of achieving this program, but there are a number of methods whose detailed study permits a meaningful characterization of signals that reconciles mathematical description and physical intuition, hopefully closing the "golden triangle" with efficient algorithms. This is also what this book is about.

One more word. This book is not intended to be a comprehensive treatise of time-frequency analysis that would cover all aspects of the field (this can be found elsewhere; see, e.g., [21], [22], [7], or [23], to name but a few). It must rather be seen as an *exploration*, a *journey* in which stops are made when needed for addressing specific issues. The construction is in no way axiomatic or linear. What is privileged is *interpretation*, at the expense of full proofs and, sometimes, full rigor.

Roadmap. The book is organized as follows. Following this Introduction, Part I is devoted to "Basics and Constraints," i.e., to a general presentation of the fundamental problems and tools that motivate the use of a time-frequency analysis. This begins in Chapter 2 with a discussion of specific signals (encountered in physics, bioacoustics, or mathematics) that are in some sense emblematic of a need for time-frequency analysis and that will be considered again at the end of the book. Chapter 3 then discusses notions of noise, in contrast with what is expected from signals as more structured objects. The Fourier description of signals in time *or* frequency is addressed in Chapter 4, with special emphasis on Gaussian waveforms for the pivotal role they play in many questions of time-frequency analysis. This is in particular the case for the uncertainty relations that put a limit on joint localization, and that are exposed under different forms in Chapter 5. Based on the previous considerations, Chapter 6 enters the core of the subject, introducing in a mostly interpretative way basic time *and* frequency representations and distributions. This offers the possibility of revisiting uncertainty in Chapter 7, from a completely time-frequency-oriented perspective. Finally, Chapter 8 discusses the key concept of (non-)stationarity, with a revisiting of time-frequency that allows for an operational definition.

Part II of this book is concerned with a more detailed exploration of the structure of time-frequency distributions in terms of "Geometry and Statistics." Chapter 9 focuses on the geometry of the spectrogram and its interpretation. This leads on to a discussion in Chapter 10 of a number of variations aimed at sharpening a spectrogram, based on ideas of reassignment, synchrosqueezing, or sparsity. Such approaches are indeed reminiscent of alternative techniques related to the so-called Hilbert–Huang Transform, which this book digresses to examine in Chapter 11. Chapter 12 comes back to the mainstream of the book, with a deeper interpretation of the spectrogram geometry in the Gaussian case, deriving spatial organizations in the plane from the structure of the reassignment vector field. Whereas the underlying construction rules apply equally to any waveform, the noise case is more specifically addressed in Chapter 13, with uncertainty revisited in terms of statistical correlation. This is detailed further in Chapter 14, which proposes a simple (randomized lattice) model for the distribution of local maxima considered as a 2D point process. Similar considerations are followed in Chapter 15 for zeros, in connection with the theory of Gaussian Analytic Functions. The importance of spectrogram zeros is stressed by the proposal of a zeros-based algorithm for time-frequency filtering, as well as by "universal" properties attached to such characteristic points. With all the techniques discussed so far at our disposal, Chapter 16 comes back to the examples of Chapter 2, elaborating on their possible time-frequency analyses and on the information that can be gained from them.

Finally, a short Conclusion is followed by a series of commented-upon links to free software tools permitting actual implementation of most of the techniques discussed in the book.

Part I

Basics and Constraints

2   Small Data Are Beautiful

The Introduction made a number of claims about the relevance of time-frequency approaches in signal processing, sketching some kind of a program for the present book. Let us start with three examples supporting those claims: one in physics, one in bioacoustics, and one in mathematics.

2.1   Gravitational Waves

The first direct observation of a gravitational wave was reported in early 2016 [5]. Since their prediction by Albert Einstein as a consequence of his theory of general relativity, proof of their existence had long been awaiting direct evidence, mostly because the extremely tiny effects they induce on matter make their detection a formidable challenge. The search for gravitational waves has therefore led to ambitious research programs based on the development of giant interferometers. The rationale is that the propagation of a gravitational wave essentially modifies the local structure of space-time, with the consequence that its impinging on an interferometer produces a differential effect on the length of its arms, and hence an oscillation in the interference pattern. Similarly to electromagnetic waves that result from accelerated charges, gravitational waves result from accelerated masses and, to be detectable, only extreme astrophysical events can be considered as candidates for producing gravitational waves. The preferred scenarios are mergers of compact binaries made of neutron stars or black holes. Within this picture of two very massive objects revolving around each other, the loss of energy due to the hypothesized radiation of a gravitational wave is expected to make them get closer and closer, hence causing them to revolve around each other at a faster and faster pace for the sake of conservation of the angular momentum. The overall result is that the signature of such a gravitational wave in the interferometric data takes the form of a "chirp" (i.e., a transient waveform modulated in both amplitude and frequency), with an increase in both amplitude and instantaneous frequency during the inspiral part that precedes the coalescence.

The event corresponding to the first detection (referred to as GW150914) was precisely of this type. It consisted of the merger of two black holes (each of about 30 solar masses), with an observation made of two transient signals of short duration (a fraction of a second), which were detected by the two LIGO interferometers located in Hanford, WA and Livingston, LA, respectively.

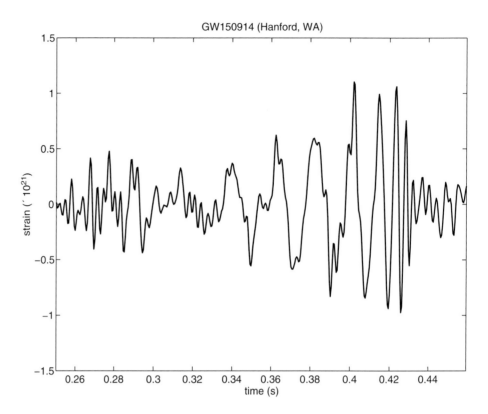

Figure 2.1 Gravitational wave chirp 1. This figure plots the temporal signature of GW150914, as it was recorded by the LIGO interferometer in Hanford, WA, and pre-processed for getting rid of known perturbations due to the measurement system.

The temporal signature of GW150914, as it was recorded by the LIGO interferometer in Hanford, WA, and pre-processed for getting rid of known perturbations due to the measurement system, is plotted in Figure 2.1. This plot gives us an idea of the chirping nature of the (noisy) waveform, but a much clearer picture of what happens is obtained when we turn to the time-frequency plane, as shown in Figure 2.2.[1]

> Gravitational waves offer an example of "small" signals (a few thousand samples at most), with characteristics that convey physical information about the system from which they originate.

Remark. Although "small," gravitational wave signals result from a prototypical example of "big science": 45 years of efforts at the forefront of technology for developing giant interferometers with arms 4 km long and a sensitivity of 10^{-21}, thousands of researchers and engineers, more than 1,000 coauthors in the landmark publication [5]... Each data point, therefore, has an immense value, calling again for accurate methods of

---

[1] In this figure, as in most figures throughout the book, time is horizontal, frequency vertical, and the energy content is coded in gray tones, ranging from white for the lower values to black for the maximum.

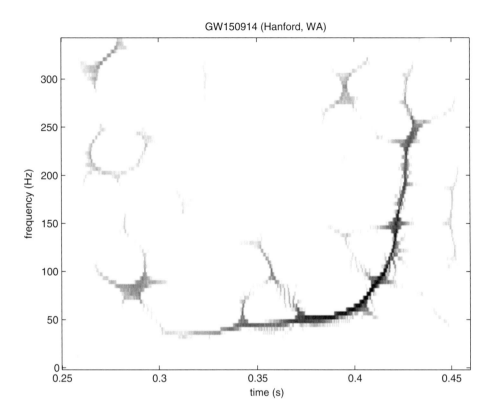

Figure 2.2 Gravitational wave chirp 2. The waveform plotted in Figure 2.1 is displayed here as an energy distribution in the time-frequency plane (namely, a "reassigned spectrogram," which will be introduced in Chapter 10). As compared to the time plot of the signal, this "musical score" reveals in a much clearer way the inner structure of the waveform, namely the frequency evolution of an ascending chirp. The energy scale is logarithmic, with a dynamic range of 24 dB.

analysis, in particular for comparing observation to theoretical models and confirming the amazing agreement that has been obtained so far [5].

Detecting gravitational waves, de-noising the corresponding chirps, and extracting physical information from them can take advantage of time-frequency approaches. We will come back to this in Chapter 16.

2.2   Bats

By following the musical score analogy outlined previously, we can switch from music to singing voice, and from singing voice to speech. All of these instances of audio signals offer countless opportunities for a time-frequency analysis aimed at displaying inner structures of sounds in a graphic, easily understandable way that matches perception. And indeed, it is not by chance that one of the first books ever published on time-frequency analysis [4] had the title *Visible Speech*!

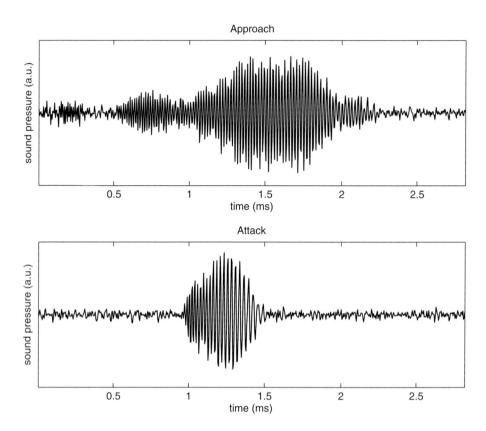

Figure 2.3 Two examples of bat echolocation calls 1. The two waveforms displayed in this figure are part of a sequence that lasts for about 1 s and contains a few dozens of such calls, with a structure (duration, spectrum, and modulations) that varies in between the beginning of the active part ("approach") and the end of the sequence ("attack").

Those common situations enter what is essentially a framework of *communication*, in which some "message" is sent by somebody, somewhere, to be received by somebody else, elsewhere. We could comment further on specific features attached to such audio signals but we will not here. We will, rather, choose as examples other types of waveforms that share much with conventional audio signals, but which differ from speech or music in at least two respects. First, whereas the transmission of a speech message can be viewed as "active" by the speaker and "passive" by the listener, there exist other situations where the system is doubly "active" in the sense that the emitter is at the same time the receiver, and where the received information is not so much the message itself as it is the modifications it may have experienced during its propagation. Second, although they are acoustic, the transmitted signals can have a frequency content that lies outside of the audio range. These two ingredients are typical of the *echolocation* system used by bats and, more generally, by other mammals such as dolphins, or even by humans in detection systems such as radar, sonar, or nondestructive evaluation.

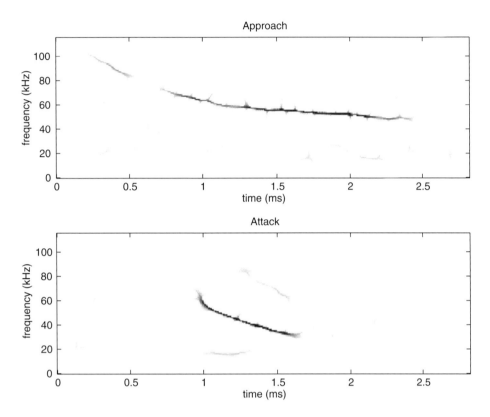

Figure 2.4 Two examples of bat echolocation calls 2. The two waveforms plotted in Figure 2.3 are displayed here as energy distributions in the time-frequency plane. As compared to the time plots of the signals, those "musical scores" reveal in a much clearer way the inner structure of the waveforms, e.g., the nature of their frequency modulations. In both diagrams, the time span is as in Figure 2.3, and the energy scale is logarithmic, with a dynamic range of 30 dB.

If we confine ourselves to bats, the story begins in 1794 when Lazzaro Spallanzani first suggested – on the basis of crucial, yet cruel experiments – that bats should have some specific sensorial capability for navigating in the dark [24]. It seemed to be related to hearing rather than to sight, since it was altered when making the animal mute and/or deaf, while making it blind was of no consequence on the flight. This question puzzled physiologists for almost two centuries, until the zoologist Donald W. Griffin reopened this mysterious case in 1938 together with the engineer George W. Pierce, who had just developed a new kind of microphone that was sensitive to ultrasounds, i.e., sounds whose frequency is above the upper limit of perception of the human ear (~ 20 kHz). In this way they were able to prove that bats were indeed emitting ultrasounds [25], and their study launched a fascinating area of research [26], with implications in both biology and engineering.

Two typical bat echolocation calls, emitted by *Myotis mystacinus* when hunting and recorded in the field, are plotted in Figure 2.3. The two waveforms are part of a sequence that lasts for about 1 s and contains a few dozen such calls, with a structure (duration, spectrum, and modulations) that varies in between the beginning of the active part (the so-called "approach" phase, during which the bat manages to get closer to the target it has identified as a potential prey item) and the end of the sequence (the "attack" phase, which terminates with the actual catch). Thanks to the time-frequency methods that will be described later in this book, we can get a much clearer picture of the inner structure of those waveforms by drawing their "musical score" as shown in Figure 2.4 (in this figure, as in Figure 2.2, we made use of a "reassigned spectrogram"). As for gravitational waves, they happen to be "chirps," with characteristics that vary within a sequence. From those diagrams, we can expect to extract more easily, and in a more directly interpretable way, the necessary information about the why and how of the observed signals in relation with a given task.

> Bat echolocation calls are an example of "small" signals (a few hundred samples at most), with a well-defined time-frequency structure whose fine characterization calls for precise analysis tools.

As for gravitational waves, we will come back to this in Chapter 16.

2.3   Riemann-Like Special Functions

The third family of "small" signals we will mention as an example is somewhat different since it concerns mathematics and, more precisely, some special functions.

Interest in the complementary descriptions of special functions beyond the mere inspection of their analytical formulation has been raised, e.g., by Michael V. Berry, who has suggested transforming such functions into sounds and listening to them [27]. The examples he chose are related to Riemann's zeta function, an analytic function of the complex variable $z \in \mathbb{C}$ which reads

$$\zeta(z) = \sum_{n=1}^{\infty} \frac{1}{n^z}, \tag{2.1}$$

and admits the equivalent representation (Euler product):

$$\zeta(z) = \prod_{p \in \mathcal{P}} \frac{1}{1 - p^{-z}}, \tag{2.2}$$

where $\mathcal{P}$ stands for the set of all prime numbers. The distribution of primes turns out to be connected with that of the zeros of the zeta function and, since a *spectrum* can be attached to this distribution [28], this paves the way for giving a real existence to a "music of primes" that had previously been evoked as an allegory.

> Of course, in parallel to the *hearing* of this music, we can naturally think of *seeing* it by writing down its "musical score" thanks again to some time-frequency analysis.

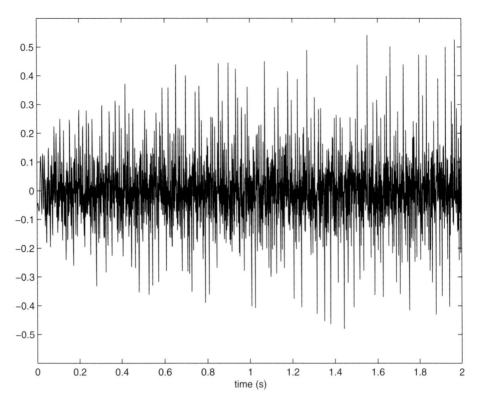

Figure 2.5 Zeta function 1. For a sampling rate fixed to 32 kHz, this figure plots the first 2 seconds of the (real part of) the waveform $Z(t)$ defined in (2.3), which essentially describes $\zeta(z)$ as a function of its imaginary part $\text{Im}\{z\} = t$ for the fixed value of its real part $\text{Re}\{z\} = 1/2$.

Following [27], we can consider the function

$$Z(t) = \zeta\left(\tfrac{1}{2} + it\right)\exp\{i\,\theta(t)\},\tag{2.3}$$

with

$$\theta(t) = \text{Im}\left\{\Gamma\left(\tfrac{1}{2}\left(\tfrac{1}{2} + it\right)\right)\right\} - (t\log\pi)/2.\tag{2.4}$$

This function is of special importance with respect to the so-called *Riemann hypothesis*, which stipulates that all zeros of the zeta function that fall within the strip $0 < \text{Re}\{z\} < 1$ are aligned along the only line given by $\text{Re}(z) = 1/2$. A plot of the first 2 seconds of the function $Z(t)$, sampled at 32 kHz [27], is given in Figure 2.5. It rather looks like noise, and gives few insights into the spectral structure, if any. In contrast, the time-frequency image given in Figure 2.6 evidences a fairly well-structured organization in terms of up-going chirps, that calls for explanations. In this case, the chosen time-frequency representation is an "ordinary" spectrogram, so as to put emphasis on the crystal-like structure of zeros in the time-frequency plane.

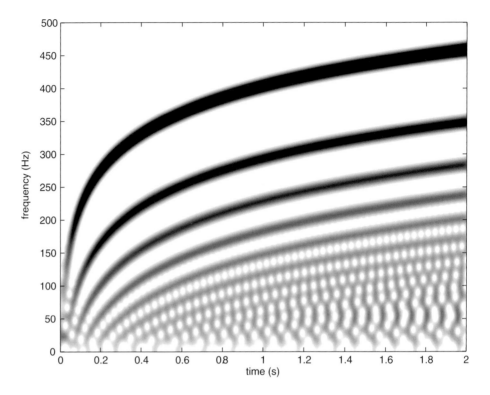

Figure 2.6 Zeta function 2. The waveform plotted in Figure 2.5 is displayed here as an energy distribution in the time-frequency plane (spectrogram). As compared to the time history of the signal, this "musical score" reveals in a much clearer way the inner structure of the waveform, namely the existence of ascending chirps. The energy scale is logarithmic, with a dynamic range of 15 dB.

As for bats and gravitational waves, we will come back to this in Chapter 16, together with some other examples of special functions (like Weierstrass's function) that admit natural time-frequency interpretations.

2.4   Chirps (Everywhere)

The three (families of) signals considered so far are just examples, and we are far from exhausting the number of situations where waveforms of a similar nature are encountered. If we consider their structure, they all share the common property of being (possibly multicomponent) "chirps." In other words, they are basically characterized by a well-defined structure that takes the form of time-frequency *trajectories* which reflect the existence of *sweeping* frequencies. This phenomenological description calls for considering what being a "chirp" means, in more mathematical terms.

The concept of frequency is indeed closely related to notions such as oscillations or cycles – which are ubiquitous in nature and technology, from the motion of celestial objects to atomic descriptions – via biological rhythms or rotating machinery. Therefore:

> Chirps appear essentially as transient time-dependent variations on sustained har-
> monic oscillations, whence their ubiquity.

In order to make this point clearer, consider a simple pendulum of length L_0 in the
usual gravity field. In the approximation of small oscillations, it is well-known that the
angle $\theta(t)$ is governed by the equation

$$\frac{d^2\theta}{dt^2}(t) + \frac{g}{L_0}\theta(t) = 0, \tag{2.5}$$

where g stands for the acceleration of gravity. Up to some pure phase term, the solution
of (2.5) is simply

$$\theta(t) = \theta_0 \cos \omega_0 t, \tag{2.6}$$

with θ_0 the maximum amplitude of the oscillation, $\omega_0 = \sqrt{g/L_0}$ its (angular) frequency,
and $T = 2\pi/\omega_0$ its period.

Let us then make the pendulum length become a slowly varying function of time (by
imposing, e.g., L_0 to be transformed into $L(t) = L_0(1 + \varepsilon t)$, with $\varepsilon > 0$ small enough to
keep the length variation small at the scale of one oscillation). This results in oscillations
that become time-dependent, with an "instantaneous" frequency that is almost "frozen"
on a short-term basis, yet in the longer term undergoes the evolution $\omega(t) \approx \sqrt{g/L(t)}$,
which is progressively slowed down as the pendulum length is increased. If we further
include viscous damping, the actual amplitude of the oscillations becomes time-varying,
with an exponential decay. To summarize, combining both effects transforms a sine
wave into a chirp!

This very simple example of a damped pendulum with time-varying length illustrates what we understand when adding "instantaneous" to words such as amplitude
or frequency. In the nominal situation of a pure harmonic oscillation $y(t)$, there is no
ambiguity in writing

$$y(t) = a \cos \omega t \tag{2.7}$$

and in considering that a is its amplitude and ω its frequency. When accepting some
possible time dependencies, it is tempting to generalize (2.7) by writing

$$x(t) = a(t) \cos \varphi(t), \tag{2.8}$$

letting a become a (nonnegative) function of time $a(t)$, and replacing ωt with a phase
term $\varphi(t)$ undergoing some possibly nonlinear evolution.

Unfortunately, there is no unique way of expressing a given observation $x(t) \in \mathbb{R}$ in a
form such as (2.8). The usual way out is to come back to (2.7) and write $x(t) = \text{Re}\{z_x(t)\}$,
i.e., to consider that $y(t)$ is the real part of some complex-valued signal $z_y(t)$, thus calling
for a decision on what the imaginary part should be. A "natural" choice is to complement
the *cosinusoidal* real part with a *sinusoidal* imaginary part, so that

$$z_y(t) = a \exp\{i\omega t\}. \tag{2.9}$$

This is the classic "Fresnel" (or "Argand") representation of a monochromatic wave, whose interpretation in the complex plane is particularly appealing. Indeed, as parameterized by time t, the complex-valued signal $z_y(t)$ in (2.9) can be seen as a rotating vector – with real and imaginary parts as coordinates – whose extremity describes a *circle* with a *constant* angular speed. The amplitude a is precisely the modulus $|z_y(t)|$, whereas the angular speed, which is the time derivative of the phase ωt, is identified with the frequency.

Given this interpretation, extending to time-varying situations is fairly obvious. It amounts to adding an imaginary part to the real-valued observation $x(t) \in \mathbb{R}$. Since there is no unicity for such an extension, the most "natural" one consists in mimicking the relationship that exists between a cosine and a sine in a complex exponential. As it can be easily established, the linear filter which turns a cosine into a sine (with the exact same amplitude and frequency) has for transfer function $H(\omega) = -i \operatorname{sgn} \omega$. In the time domain, this corresponds to the *Hilbert transform* $\mathbf{H}$ such that:

$$(\mathbf{H}x)(t) = \frac{1}{\pi} \text{p.v.} \int_{-\infty}^{\infty} \frac{x(s)}{t-s} \, ds, \tag{2.10}$$

where "p.v." indicates that the integral has to be computed as a "principal value" in Cauchy's sense, i.e., as

$$\text{p.v.} \int_{-\infty}^{\infty} = \lim_{\epsilon \to 0} \left[\int_{-\infty}^{-\epsilon} + \int_{\epsilon}^{+\infty} \right]. \tag{2.11}$$

Applying the above recipe for the complexification of the real-valued signal $x(t)$ thus leads to the well-known solution of the so-called *analytic signal* which reads [29, 30]:

$$z_x(t) = x(t) + i\,(\mathbf{H}x)(t). \tag{2.12}$$

The "real/imaginary" expression (2.12) admits the equivalent "modulus/phase" representation:

$$z_x(t) = a(t) \, \exp\{i\varphi(t)\} \tag{2.13}$$

which is now unique and from which an "instantaneous amplitude" $a(t) \geq 0$ and an "instantaneous frequency" $\omega(t)$ can be derived as in the purely harmonic case, *mutatis mutandis*:

$$a(t) = |z_x(t)| \quad ; \quad \omega(t) = \frac{d}{dt}\arg\{z_x(t)\}. \tag{2.14}$$

Generalizing upon the pendulum example, "slowly-varying" conditions are often assumed for chirps. Usual heuristic conditions assume that $|\dot{a}(t)/a(t)| \ll |\dot{\varphi}(t)|$, i.e., that the amplitude is *almost constant* at the scale of one pseudo-period $T(t) = 2\pi/|\dot{\varphi}(t)|$, and that $|\ddot{\varphi}(t)/\dot{\varphi}^2(t)| \ll 1$, i.e., that the pseudo-period $T(t)$ is itself *slowly varying* from one oscillation to the next [31, 32].

Remark. Although "classic" and the most used one, the definition of "instantaneous frequency" on the preceding page may lead to some paradoxes and present difficulties in its physical interpretation. It can, for instance, attain negative values, or have excursions outside of a frequency band in which the signal spectrum is supposed to be limited [22].

One can argue, however, that such unexpected behaviors apply to situations that can be considered as departing from the assumed model. This is especially the case when more than one chirp component is present at a given time, a situation in which one would normally expect two values; of course, this is impossible with the definition of a mono-valued function. Without having recourse to time-frequency analysis (which will prove to be a better approach), alternative definitions – with their own pros and cons – have been proposed. These will not be discussed here; for more information, refer to [33].

In retrospect, it is clear that the examples considered in Sections 2.1–2.3 of this chapter can be reasonably considered as *multicomponent chirps* (also known as *multicomponent AM-FM (Amplitude Modulated – Frequency Modulated) signals*), all of which accept a model of the form

$$x(t) = \sum_{k=1}^{K} a_k(t) \cos \varphi_k(t). \tag{2.15}$$

We confine ourselves here to (and will discuss further in Chapter 16) a few such signals, but it is worth stressing that waveforms of a very similar nature can be found in many domains. The following list shows a few of the possibilities one can mention:

- *Birdsongs* – This is of course the first instance, and the one from which the name "chirp" comes from since, according to Webster's 1913 Dictionary, a chirp is "a sharp sound made by small birds or insects."
- *Animal vocalizations* – Besides birds (and bats), many animals make use of transient chirps or longer AM-FM signals, mostly for communication purposes: short chirps by frogs, longer vocalizations by whales (listen, e.g., to the many sound files available at http://cis.whoi.edu/science/B/whalesounds/index.cfm).
- *Audio, speech, and music* – Audio is a rich source of chirping waveforms: singing voice, vibrato, glissando . . . Several examples have been exhibited since the early days of time-frequency analysis [4].
- *Dispersive media* – A brief pulse can be idealized as the coherent superposition of "all" possible frequencies. If such a waveform is sent through a dispersive medium for which the group velocity is frequency-dependent, the different frequency components travel at different speeds, resulting in a distortion that, in time, spreads the highly localized pulse and transforms it into a chirp. This can be observed, e.g., in the backscattering from simply shaped elastic objects in underwater acoustics [34].
- *Whistling atmospherics* – A companion example where chirps are due to dispersion is to found in geophysics, with low-frequency "whistlers" that can follow impulsive atmospherics (such as lightning strokes) after propagation in the outer ionosphere [35].
- *Turbulence* – Turbulent flows can be given both statistical and geometrical descriptions. In 2D, disorder can be modeled *via* some random distribution of spiraling coherent structures, namely vortices (e.g., the swirls that can be observed in rivers downstream from a bridge). Intersecting such a 2D vortex

results in a 1D profile taking the form of a "singular" core surrounded by a chirp [36].

- *Warping* – In a natural generalization of the pendulum example, oscillations of moving sources lead to chirping behaviors. This was noticed for gravitational waves, but this also applies to Doppler effect, where a pure tone emitted by a moving source in its own referential ends up with a modulated wave that is perceived by a receiver as compressed or stretched when it passes by. Similarly, acceleration warps the characteristic rotation frequencies of an engine.
- *Electrophysiological signals* – When recording electroencephalograpic (EEG) signals to monitor brain activity, it turns out that the abnormal neural synchrony attached to epilepsy has a chirp signature [37]. In a different context, uterine electromyographic (EMG) signals do chirp too during contraction episodes, with a bell-shaped ascending/descending narrowband frequency component [38].
- *Critical phenomena* – Oscillations that accelerate when approaching a singularity are in some sense "universal" [39]. This has been advocated for identifying precursors in domains as diverse as earthquakes or financial crashes.
- *Man-made sounders* – Mimicking the echolocation principle used by bats or dolphins, some human-made systems make use of chirp-like signals for sounding their environment. One can cite the field of "fisheries acoustics" where FM signals are used by broadband sounders [40]. Another key example is vibroseismics, where sweeping frequencies are sent through the ground in a particular area – for the sake of oil exploration – by means of specially equipped trucks [41].

We could provide many more such examples, but we will stop here. One more good reason for closing the list with vibroseismics is that it is emblematic of the way science can be driven by applied problems. Indeed, it is no exaggeration to say that the whole field of wavelet analysis – which can be viewed as a twin sister to the time-frequency analysis on which this book is primarily focused – originated from Morlet's concerns to improve upon "classic" Fourier-based vibroseismic signal analysis. When based on solid mathematical grounds and equipped with efficient algorithms, the roads Morlet started to explore led to unexpected "hot spots" with new perspectives far beyond the initial problem and its original time-frequency flavor.

3   Of Signals and Noise

In nearly all real-life situations, we observe that measurements are made of a mixture of "signal" and "noise." The term "signal" is generally used to refer to a waveform that carries some information that is considered to be meaningful to the receiver, while "noise" refers to any kind of disturbance – both from the measuring device itself and from other sources – that blurs the signal of interest.

3.1   Order versus Disorder

Denoising a signal, i.e., disentangling it from noise, is a primary objective of signal processing and, for this process to make sense, some assumptions must be made about both signals and noise.

As we have noted previously, whether a signal is considered "of interest" is often a matter of subjective appreciation. As long as the distinction "signal versus noise" is considered, we will make use of the intuition we have gained from the examples in Chapter 2 and adopt the following definition of "signal" for the purposes of this book:

> A *signal* will be characterized as a waveform having some well-defined structured organization in the time-frequency plane.

This is typically the case of the multicomponent chirps considered so far as examples, with energy ribbons along time-frequency trajectories.

In contrast to the concept of a signal, which is associated with *coherence* in structure and organization, noise is considered more *erratic*, with structures that can hardly be described as established and coherent.

> In other words, one can think of the conceptual separation between signal and noise as an opposition between *order* and *disorder*.

It is worth mentioning that this opposition suggests an implicit line of reasoning that will be followed throughout this book, intertwining ideas from both *geometry* (because assuming a coherent time-freqency structure for signals leads to questions of spatial organization in the plane) and *statistics* (because the concept of disorder – if we forget, in a first approach, deterministic models stemming from, e.g., chaos theory – is intimately linked with an assumption of randomness).

Let us just mention at this point that this text will prove that both perspectives of geometry and statistics share common grounds that allow for unified interpretations and uses: this will be explored further in Part II.

3.2   Signals

At a very pragmatic level, and if we stay at the level of data and not of models, any observation we can make of a phenomenon that gives rise to some recorded signal is of *finite duration*: it must have started at some time in a finite past, and it terminates at best at the present time, with no possible extension to some infinite future. When considered in terms of amplitude and/or power, physics and engineering also dictate that *no infinite values are allowed*. The overall result is that convenient frameworks for describing real-life signals are the function spaces $L^p(\mathbb{R})$ ($p \in [1, +\infty)$), defined by:

$$L^p(\mathbb{R}) = \left\{ x(t) \in \mathbb{C} \,|\, \|x\|_p = \left(\int_{-\infty}^{\infty} |x(t)|^p \, dt \right)^{1/p} < +\infty \right\}. \tag{3.1}$$

Choosing, e.g., $p = 1$ or $p = 2$ corresponds to the standard cases of integrable or square-integrable signals, respectively. A simple physical interpretation is attached to $L^2(\mathbb{R})$ since square-integrability means *finite energy*. Of particular interest is the case $L^1(\mathbb{R}) \cap L^2(\mathbb{R})$, for which it is well-known that a powerful tool for handling the corresponding signals is the *Fourier transform* [43] which determines the frequency spectrum $X(\omega)$ of a given signal $x(t)$ according to:

$$X(\omega) = \int_{-\infty}^{\infty} x(t) \, \exp\{-i\omega t\} \, dt, \tag{3.2}$$

with the inversion formula:

$$x(t) = \int_{-\infty}^{\infty} X(\omega) \, \exp\{i\omega t\} \, \frac{d\omega}{2\pi}. \tag{3.3}$$

The Fourier transform is an isometry of $L^2(\mathbb{R})$ or, in other words, it is such that it satisfies Plancherel's identity:

$$\langle x, y \rangle = \int_{-\infty}^{\infty} x(t) \, y^*(t) \, dt = \int_{-\infty}^{\infty} X(\omega) \, Y^*(\omega) \, \frac{d\omega}{2\pi}, \tag{3.4}$$

with the consequence that the signal energy E_x can be equivalently expressed by summing either $|x(t)|^2$ (the time energy density, or instantaneous power) or $|X(\omega)|^2$ (the frequency (or spectral) energy density); this is Parseval's identity, which reads:

$$E_x = \int_{-\infty}^{\infty} |x(t)|^2 \, dt = \int_{-\infty}^{\infty} |X(\omega)|^2 \, \frac{d\omega}{2\pi}. \tag{3.5}$$

Remark. Introducing the notation $e_\omega(t) = \exp\{i\omega t\}$, the analysis/synthesis relations (3.2) and (3.3) can be rewritten as:

$$X(\omega) = \langle x, e_\omega \rangle; \tag{3.6}$$

$$x(t) = \int_{-\infty}^{+\infty} \langle x, e_\omega \rangle\, e_\omega(t)\, \frac{d\omega}{2\pi}. \tag{3.7}$$

As for (3.6), this writing simplifies the interpretation of the Fourier transform as the "projection" of a signal onto the family of complex exponentials, i.e., monochromatic waves. Those complex exponentials serve in turn as building blocks for recovering the original signal thanks to the properly weighted superposition given in the reconstruction formula (3.7). While this makes perfect sense from a mathematics perspective, the link between physics and mathematics in the "golden triangle" of Figure 1.1 can present something of a paradox when it comes to interpretation. Indeed, the expansion (3.7) states that a finite energy signal results from the infinite superposition of the basic waveforms $e_\omega(t)$, none of which is of finite energy! This is an example of the discrepancy that might exist between physical interpretation and mathematical formalization. We will see later how a joint time-frequency approach allows us to work through this apparent contradiction, and to better reconcile the differences between physics and mathematics.

Whereas $L^2(\mathbb{R})$ is a particularly convenient setting for most current signals when we want to be close to real-world data, numerous other options are offered for generalized idealizations and mathematical models, although some of these may have to exist in theory only.

Example 1. A first example of idealization is given by the aforementioned complex exponentials $e_\omega(t)$, which do not exist *per se* in nature but may serve as a convenient model allowing for easier mathematical manipulations. Such monochromatic waves have infinite energy, but they have finite power, – i.e., they are one instance of infinite duration signals such that:

$$\lim_{T\to\infty} \frac{1}{T} \int_{-T/2}^{+T/2} |x(t)|^2\, dt < +\infty. \tag{3.8}$$

Example 2. A second example is provided by the indicator function:

$$\text{rect}_T(t) = \frac{1}{T}\, 1_T(t), \tag{3.9}$$

which attains the value $1/T$ on the interval $[-T/2, +T/2]$, is identically zero elsewhere, and has as its Fourier transform the quantity

$$\text{Rect}_T(\omega) = \frac{\sin \omega T/2}{\omega T/2}. \tag{3.10}$$

The signal (3.9) is integrable and of unit area whatever the duration T (provided the value of T is finite and nonzero). The case where T goes to zero is of special interest, since it can be argued that this models the idealization of a perfectly localized impulse function. This limit (which is no longer square-integrable) formally defines the so-called Dirac's "δ-function"

$$\delta(t) = \lim_{T \to 0} \text{rect}_T(t), \tag{3.11}$$

which is actually not a function, but a distribution. To make things more precise (see, e.g., Appendix A.6 in [20]), this quantity should be defined only via its action on well-behaved test-functions $\varphi(.)$ according to:

$$\langle \varphi, \delta \rangle = \int_{-\infty}^{+\infty} \varphi(u)\, \delta(u)\, du = \varphi(0), \tag{3.12}$$

but we will here adopt its loose usage, favoring interpretation at the expense of rigor. This will allow for simple symbolic calculations such as pointwise evaluations according to:

$$\int_{-\infty}^{\infty} x(t)\, \delta(t - s)\, dt = x(s), \tag{3.13}$$

as well as for using (in many circumstances) the Fourier duality of the pair "Dirac/unity" as a result of the formal limit:

$$\lim_{T \to 0} \text{Rect}_T(\omega) = 1. \tag{3.14}$$

3.3   Noise

The idea of disorder suggests that noise has a random nature and that, given an observation, the noise part is only one realization of some stochastic process. It thus follows that the analysis usually has to be inferred only on the basis of such an effective realization, hopefully complemented by some *ensemble* knowledge of the statistical properties of the underlying process.

We will here stick to the simplest type of noise (which can be considered as representative of the most disordered situation, and from which more structured situations can be derived through filtering operations) – namely, *white Gaussian noise*.

The rationale behind the introduction of white (Gaussian) noise is quite clear. It relies on the analogy with the spectrum of white light, as it can be observed with a prism, and states that "whiteness" just reflects the equipartition of all wavelengths (or colours). Being more precise, however, requires some care.

Real-valued white Gaussian noise — One first observation is that the idea of "maximum disorder" attached to white noise should apply not only to frequencies (with all values equally represented) but also to times, so that equal randomness would occur at any instant. This situation of sustained randomness, with no privileged time reference, makes white noise $n(t)$ enter the framework of *stationary processes*. If we confine ourselves to first- and second-order properties, stationarity of a process $x(t) \in \mathbb{R}$ requires that a constant m_x and a function $\gamma_x(\tau)$ exist such that [42, 43]

$$\mathbb{E}\{x(t)\} = m_x \tag{3.15}$$

(where $\mathbb{E}\{.\}$ stands for the expectation operator) and

$$\text{cov}\{x(t), x(t - \tau)\} = \mathbb{E}\{[x(t) - m_x][x(t - \tau) - m_x]\} = \gamma_x(\tau). \tag{3.16}$$

In other words, a (second-order) stationary process has a constant mean and a covariance function that only depends on the time lag between the two considered instants.

Remark. For the sake of simplicity, and without any loss of generality, we can (and often do) assume that $m_x = 0$. In the most general case of stationarity, independence with respect to some absolute reference time goes beyond second-order and concerns all moments. In the Gaussian case, however, the first and second moments are sufficient to fully characterize the distribution, with the consequence that second-order stationarity implies full stationarity.

By construction, a stationary process has a constant variance, namely

$$\text{var}\{x(t)\} = \gamma_x(0) \tag{3.17}$$

and the trajectories of its realizations cannot be of finite energy. As for complex exponentials, envisioning their Fourier transform has to be made in a generalized sense, which simply prohibits defining a power spectrum by computing an ordinary Fourier transform [44]. A way out is possible, however, which parallels in some sense what is done when computing the spectrum of a square-integrable signal. Indeed, we have seen with (3.5) that the spectral energy density of a finite energy signal $x(t)$ is directly given by squaring its Fourier transform, but a straightforward computation shows that the very same quantity can be expressed as:

$$|X(\omega)|^2 = \int_{-\infty}^{\infty} \tilde{\gamma}_x(\tau) \exp\{-i\omega\tau\} \, d\tau, \tag{3.18}$$

where

$$\tilde{\gamma}_x(\tau) = \int_{-\infty}^{\infty} x(t) \, x(t - \tau) \, dt \tag{3.19}$$

is the deterministic correlation function of $x(t)$. The stochastic counterpart of this result is the Wiener-Khintchine-Bochner theorem [42] which states that the power spectrum density $\Gamma_x(\omega)$ of a stationary process $x(t)$ is:

$$\Gamma_x(\omega) = \int_{-\infty}^{\infty} \gamma_x(\tau) \exp\{-i\omega\tau\} \, d\tau, \tag{3.20}$$

i.e., precisely the Fourier transform of the stationary covariance function $\gamma_x(\tau)$.

Remark. The parallel between the deterministic and stochastic cases can be further demonstrated by the observation that, in both cases, the correlation function is a measure of similarity between a signal and its shifted templates, expressed by an appropriate inner product. Whereas the finite energy case relies on the ordinary inner product of $L^2(\mathbb{R})$:

$$\langle x, y \rangle_{L^2(\mathbb{R})} = \int_{-\infty}^{\infty} x(t)\, y(t)\, dt, \qquad (3.21)$$

an alternative is offered in the stochastic case by the covariance:

$$\langle x, y \rangle_{\mathcal{L}^2(\mathbb{R})} = \mathbb{E}\{x(t)\, y(t)\} \qquad (3.22)$$

which also defines an inner product, in the space $\mathcal{L}^2(\mathbb{R})$ of finite power stationary processes. In both cases, the correlation function results from a comparison between $x(t)$ and $y(t) = (\mathbf{T}_\tau x)(t)$, where $\mathbf{T}_\tau$ stands for the time shift operator such that $(\mathbf{T}_\tau x)(t) = x(t-\tau)$. This results in a common measure of the form $\langle x, \mathbf{T}_\tau x \rangle$, to be Fourier transformed upon the lag variable τ.

Coming back to white noise $n(t)$, its spectrum has been said to be flat, implying that its Fourier transform is infinitely peaked:

$$\gamma_n(\tau) = \gamma_0\, \delta(\tau). \qquad (3.23)$$

In terms of (time) support, this infinitely peaked function is the limit idealization of what physicists refer to as a "microscopic correlation." As previously explained, it goes with a diverging amplitude at lag $\tau = 0$, which corresponds to an undefined variance. This pathological situation (which is typical of the continuous-time setting considered so far) can, however, be regularized when discretizing by projecting onto an orthonormal system $\{\varphi_k(t), k \in \mathbb{N}\, \text{or}\, \mathbb{Z}\}$. Let $n_k = \langle n, \varphi_k \rangle$ be the discrete random variables obtained by projecting over this system $n(t)$, assumed to be zero-mean and satisfy (3.23). It readily follows that $\mathbb{E}\{n_k\} = 0$ and $\mathbb{E}\{n_k\, n_{k'}\} = \gamma_0\, \delta_{kk'}$, with $\delta_{kk'}$ the Kronecker symbol whose value is 1 when $k = k'$ and 0 otherwise. Linear filtering being known to preserve Gaussianity, the variables n_k are Gaussian whenever the process $n(t)$ is. To summarize:

The projection of a continuous white Gaussian noise $n(t) \in \mathbb{R}$ onto an orthonormal system results in a collection of uncorrelated Gaussian random variables..

Please note that such a projection can be achieved in different ways, either in the spirit of sampling with $\varphi(t) = \mathbf{1}_{\Delta t}(t - k\, \Delta t)/\Delta t^{-1/2}$, or by using more elaborate systems such as Hermite's, as will be done later.

Complex-valued white Gaussian noises — Although in many cases it is natural to assume that noise is real-valued, there might be reasons to consider it as complex-valued. These reasons can include the nature of the observation (think of bivariate – e.g., polarimetric – data whose processing can be more effective in the complex plane), because complexifying real-valued data may ease post-processing (think of making signals analytic so as to have access to instantaneous frequency), or simply because such a model, even if somehow disconnected from reality, permits calculations that are easier to manipulate, while offering a good approximation to actual situations.

Let us write $x(t) \in \mathbb{C}$ as $x(t) = x_r(t) + i x_i(t)$ and assume the process to be zero-mean (i.e., that $\mathbb{E}\{x_r(t)\} = \mathbb{E}\{x_i(t)\} = 0$, with $x_r(t)$ and $x_i(t)$ jointly stationary). Second-order properties now involve statistical dependencies between real and imaginary parts, which can be expressed by means of a *correlation* function $\gamma_x(t)$ and a *relation* function $r_x(t)$, defined respectively by:

$$\gamma_x(\tau) = \mathbb{E}\{x(t) x^*(t - \tau)\}; \tag{3.24}$$
$$r_x(\tau) = \mathbb{E}\{x(t) x(t - \tau)\}. \tag{3.25}$$

An explicit evaluation of those quantities leads to:

$$\gamma_x(\tau) = [\gamma_{x_r}(\tau) + \gamma_{x_i}(\tau)] + i [\eta(\tau) - \eta(-\tau)] \tag{3.26}$$
$$r_x(\tau) = [\gamma_{x_r}(\tau) - \gamma_{x_i}(\tau)] + i [\eta(\tau) + \eta(-\tau)], \tag{3.27}$$

with $\eta(\tau) = \mathbb{E}\{x_i(t) x_r(t - \tau)\}$. Therefore, (3.26) and (3.27) completely characterize the process in the Gaussian case.

A stationary process is then said to be *proper* if its relation function is zero [45], i.e., if $\gamma_{x_r}(\tau) = \gamma_{x_i}(\tau)$ and $\eta(-\tau) = -\eta(\tau)$. It thus follows that the correlation function for a proper process is:

$$\gamma_x(\tau) = 2[\gamma_{x_r}(\tau) + i \eta(\tau)]. \tag{3.28}$$

In the Gaussian case, this is equivalent to being *circular* [45], which is the terminology that we will adopt from now on.

A further simplification occurs if we also assume that the real and imaginary parts are uncorrelated. This results in $\eta(\tau) = 0$ and, hence, $\gamma_x(\tau) = 2\gamma_{x_r}(\tau)$. Finally, the whiteness of the process may be expressed as $\gamma_{x_r}(\tau) = (\gamma_0/2) \delta(\tau)$, resulting in the model of *complex white Gaussian noise* which is characterized by the set of first and second order properties:

$$\mathbb{E}\{n(t)\} = 0; \tag{3.29}$$
$$\mathbb{E}\{n(t)n(t - \tau)\} = 0; \tag{3.30}$$
$$\mathbb{E}\{n(t)n^*(t - \tau)\} = \gamma_0 \delta(\tau). \tag{3.31}$$

As we did in the real-valued case, projecting complex white Gaussian noise onto a set of orthonormal functions $\varphi_k(t)$ results in Gaussian random variables n_k that inherit correlation/decorrelation properties from the continuous-time case, namely:

$$\mathbb{E}\{n_k\} = 0; \tag{3.32}$$
$$\mathbb{E}\{n_k n_{k'}\} = 0; \tag{3.33}$$
$$\mathbb{E}\{n_k n_{k'}^*\} = \gamma_0 \delta_{kk'}, \tag{3.34}$$

Remark. As discussed in Chapter 2, a conventional way of making a real-valued signal $x(t)$ complex is to compute its associated *analytic signal*, with the Hilbert transform $(\mathbf{H}x)(t)$ as imaginary part. In the circular case, we simply get $\eta(\tau) = (\mathbf{H}\gamma_{x_r})(\tau)$, a quantity which is such that $\eta(0) = 0$ but which does not guarantee that nonzero lags will be uncorrelated, even in the white noise case for which $\eta(\tau) = \gamma_0/2\pi\tau$: analytic white Gaussian noise, while circular, does not boil down to the complex white Gaussian noise defined in (3.32)–(3.34). We will come back to such heuristic considerations in Chapter 13, referring to [46] for a more rigorous treatment.

4   On Time, Frequency, and Gauss

Carl-Friedrich Gauss (1777–1855) was a scientific giant. Mathematician, physicist, astronomer, he contributed in an extraordinary way to fields as diverse as number theory, differential geometry or statistics, to name but a few. The breadth of his work was so great that it made inroads into disciplines (such as signal processing) that did not even exist at that time. One striking example is his discovery of what came to be called the Fast Fourier Transform algorithm in 1805, i.e., even before the actual formalization by Fourier himself (in 1811) of the trigonometric series that are now referred to as Fourier series. This work, which was not published during his lifetime, appeared only in his collected works in 1866 and it remained almost unnoticed until the 1980s (see [47] for the complete story), but it underlines in an illuminating way Gauss' interest in algorithmic efficiency.

The other well-known example to which the name of Gauss is attached is of course the *Gaussian distribution* that is of major importance in statistics and in most applied sciences. The Gaussian distribution is extensively used for two main reasons: it has a simple physical justification (central limit theorem) and its form allows for simple mathematical manipulations. Besides the statistical basis on which the Gaussian distribution is constructed, it can also be viewed as a function with its emblematic "bell-shaped" profile. From this perspective too, it is remarkable that Gaussian functions exhibit some universal properties that distinguish them from other functions in the time-frequency context that interests us here. We will see why and how in many places in this book, and this justifies starting by considering Gaussian functions in some detail.

4.1   Gauss

Let us start from the beginning and define a "Gaussian" waveform (i.e., a function in the time domain) as the following 2-parameter, 1D function of the time variable $t \in \mathbb{R}$:

$$g_{\alpha\beta}(t) = \beta \exp\left\{-\alpha t^2\right\}, \alpha > 0, \beta > 0 \qquad (4.1)$$

in which β fixes the maximum height of the function, while α determines its width. For the proper normalization $\beta = \sqrt{\alpha/\pi}$, the area under the graph turns out to be unity, and (4.1) defines the "probability density function" of some Gaussian random variable, with variance given by $\sigma^2 = 1/2\alpha$. The "bell-shaped" profile of the Gaussian function, as well as its dependence on α, is illustrated in Figure 4.1.

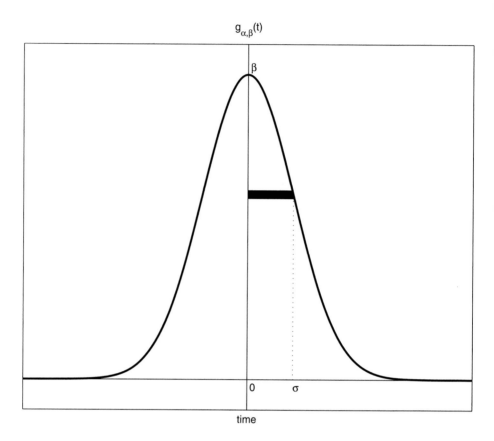

Figure 4.1 Gaussian waveform $g_{\alpha,\beta}(t)$. The height of this function is given by β, while its spread is controlled by $\sigma = 1/\sqrt{2\alpha}$ (thick line).

The smaller the value of α, the larger the value of σ and, hence, the greater the spread of the Gaussian waveform; conversely, the larger the value of α, the smaller the essential support of the graph.

Provided that β is finite, we are assured that $g_{\alpha,\beta}(t) \in L^1(\mathbb{R})$, with the result that its Fourier transform

$$G_{\alpha,\beta}(\omega) = \int_{-\infty}^{+\infty} g_{\alpha,\beta}(t) \exp\{-i\omega t\}\, dt \qquad (4.2)$$

is well-defined. The striking (yet well-known) result is that

$$G_{\alpha,\beta}(\omega) = \beta \sqrt{\frac{\pi}{\alpha}} \exp\{-\omega^2/4\alpha\}, \qquad (4.3)$$

i.e., that the Fourier transform of the bell-shaped waveform (4.1) is itself identically bell-shaped.

Remark. Gaussians are the simplest examples of a "self-Fourier" function, but they are not the only ones. One convincing point in favor of this fact is the observation that the Fourier transform, when seen as an operator, has Hermite functions $\{h_n(t), n = 0, \ldots\}$ for eigenfunctions, with eigenvalues proportional to $(-i)^n$ [48]. The first Hermite function, $h_0(t)$, identifies with the Gaussian function, thus leading back to the result obtained by a direct calculation. It follows, however, that any linear combination of Hermite functions of orders $\{n = 4m, m \in \mathbb{N}\}$ will also be self-Fourier.

4.2   From Gauss to Fourier

While the shape invariance always holds for the Gaussian function under the Fourier transformation, different normalizations (i.e., different choices for β in (4.1)) can be envisioned. For instance, forcing the area under $G_{\alpha,\beta}(\omega)$ – instead of $g_{\alpha,\beta}(t)$ – to be equal to 1, we end up with the only choice $\beta = 1$, regardless of the value of α. This in turn allows us to consider limiting cases depending upon the value of α, and it follows in particular that

$$\lim_{\alpha \to 0} G_{\alpha,1}(\omega) = \delta(\omega) \Leftrightarrow \lim_{\alpha \to 0} g_{\alpha,1}(t) = 1, \tag{4.4}$$

where $\delta(.)$ stands for the so-called "δ-function."

If we now consider a modulated version of the basic Gaussian waveform, namely

$$h_{\alpha,\beta;\omega_0}(t) = g_{\alpha,\beta}(t) \exp\{i\omega_0 t\}, \tag{4.5}$$

it readily follows from Fourier's shift theorem that

$$H_{\alpha,\beta;\omega_0}(\omega) = G_{\alpha,\beta}(\omega - \omega_0) \tag{4.6}$$

and, thus, that

$$\lim_{\alpha \to 0} H_{\alpha,1;\omega_0}(\omega) = \delta(\omega - \omega_0) \Leftrightarrow \lim_{\alpha \to 0} h_{\alpha,1;\omega_0}(t) = \exp\{i\omega_0 t\}. \tag{4.7}$$

> We recognize in this extreme case the pure tones that serve as elementary building blocks for the Fourier transform.

4.3   From Gauss to Shannon-Nyquist

According to the derivation described in Section 4.2, Fourier "modes," which consist in monochromatic waves of infinite duration, appear as the limiting case of a modulated Gaussian whose spread in time goes to infinity while its bandwidth goes to zero. Of course, one could achieve a similar result by exchanging time and frequency or, in other words, by considering a limit in which the spread in time goes to zero while bandwidth goes to infinity. This companion situation is based on time-shifted versions of the basic Gaussian waveform, namely

$$\tilde{h}_{\alpha,\beta;t_0}(t) = g_{\alpha,\beta}(t - t_0), \tag{4.8}$$

together with α going to infinity. Now, by forcing the area under this graph to have a value of 1 (as we did previously in the frequency domain), we end up with $\beta = \sqrt{\alpha/\pi}$ and, since we have

$$\tilde{H}_{\alpha,\beta;t_0}(\omega) = G_{\alpha,\beta}(\omega) \exp\{-i\omega t_0\}, \tag{4.9}$$

it readily follows that

$$\lim_{\alpha\to\infty} \tilde{H}_{\alpha,\sqrt{\alpha/\pi};t_0}(\omega) = \exp\{-i\omega t_0\} \Leftrightarrow \lim_{\alpha\to\infty} \tilde{h}_{\alpha,\sqrt{\alpha/\pi};t_0}(t) = \delta(t - t_0). \tag{4.10}$$

Introducing the notation

$$\delta_s(t) = \delta(t - s), \tag{4.11}$$

we just recover the "natural" representation in time which parallels the Fourier pair (3.2), since it reads

$$x(s) = \langle x, \delta_s \rangle; \tag{4.12}$$

$$x(t) = \int_{-\infty}^{+\infty} \langle x, \delta_s \rangle \, \delta_s(t) \, ds. \tag{4.13}$$

This representation based on δ-functions could be referred to as a Shannon-Nyquist representation.

> The reason is that it paves the way to time sampling in the case of a band-limited signal, replacing the continuum of δ-functions (required by a possibly infinite bandwidth) by a comb whose period is fixed by the highest frequency of the spectrum [43].

4.4   From Gauss to Gabor

In cases of both extremes (i.e., where α goes either to zero or to infinity), projecting a temporal signal $x(t)$ onto the corresponding families of elementary waveforms (4.7) and (4.10) results in a representation that either depends on frequency only (its Fourier spectrum, with no reference to time) or just remains unchanged (its "natural" representation, with no explicit frequency dependence).

> One very interesting thing about Gaussians with $0 < \alpha < \infty$ is that they permit a continuous transition between these two extremes, offering a description that simultaneously takes into account time and frequency dependencies.

Indeed, we can think of combining the time and frequency shifts that occur in (4.8) and (4.5), respectively, by introducing the doubly modulated waveform:

$$\psi_{\alpha,\beta;\tau,\xi}(t) = g_{\alpha,\beta}(t-\tau)\exp\{i\xi(t-\tau/2)\}. \tag{4.14}$$

Remark. The introduction of the pure phase factor $i\xi\tau/2$ in definition (4.14) is somehow arbitrary. It could be replaced *mutatis mutandis* by another convention or even be omitted, as it is often the case in the literature. The reason why it is adopted here is its ability to simplify the connection with the Bargmann transform that will be discussed later in Chapter 12; this also allows for a more symmetric treatment of the time and frequency variables.

Projecting a signal $x(t)$ over the family of elementary waveforms (4.14) gives rise to a new representation

$$\chi_{\alpha,\beta}(t,\omega) = \langle x, \psi_{\alpha,\beta;t,\omega}\rangle \tag{4.15}$$

that now jointly depends on time and frequency. Using the reconstruction formulæ (3.2) and (4.3), it can be easily established that

$$x(t) = \int\int_{-\infty}^{\infty} \langle x, \psi_{\alpha,\beta;\tau,\xi}\rangle \, \psi_{\alpha,\beta;\tau,\xi}(t) \, d\tau \frac{d\xi}{2\pi}, \tag{4.16}$$

provided that the analyzing functions in formula (4.14) satisfy the so-called admissibility condition

$$\int_{-\infty}^{\infty} |\psi_{\alpha,\beta;\tau,\xi}(t)|^2 \, dt = 1, \tag{4.17}$$

leading in turn to the normalization $\beta = (2\alpha/\pi)^{1/4}$.

Remark. This overcomplete representation is the continuous counterpart of the discrete expansion of a signal over elementary Gaussians that was proposed by Dennis Gabor in 1946 [29]. It is analogous to the quantum-mechanical representation introduced by Roy J. Glauber in 1963 [49], in which the elementary Gaussian – as first eigenstate of the harmonic oscillator – is referred to as a "coherent state." Although it is now standard material in signal processing, its first use in this field was by Carl W. Helström in 1968 [50].

When expliciting the inner product (4.15), we see that

$$\chi_{\alpha,\beta}(t,\omega) = \int_{-\infty}^{\infty} x(s) \, g_{\alpha,\beta}(s-t) \, \exp\{-i\omega(s-t/2)\} \, ds, \tag{4.18}$$

i.e., that the time-frequency representation $\chi_{\alpha,\beta}(t,\omega)$ is just a special case of a *Short-Time Fourier Transform* (STFT), with a Gaussian as short-time window.

For finite energy signals $x(t) \in L^2(\mathbb{R})$, the isometry of the Fourier transform carries over to the SFTF according to

$$\int\int_{-\infty}^{\infty} |\chi_{\alpha,\beta}(t,\omega)|^2 \, dt \frac{d\omega}{2\pi} = \|x\|_2^2. \tag{4.19}$$

The squared magnitude $|\chi_{\alpha,\beta}(t,\omega)|^2$ of the STFT appears therefore as an energy density which is classically referred to a *spectrogram*. For sake of simplicity, the specific spectrogram obtained with a Gaussian window will henceforth be referred to as a "Gaussian spectrogram." Note that, in the physics literature, the very same quantity (with position and momentum in place of time and frequency) is called a "Husimi distribution" [51].

5   Uncertainty

The time and frequency variables cannot be treated independently. They are coupled via the Fourier transform (they are sometimes said to be "canonically conjugated"). This coupling induces constrained relationships for time-frequency relationships: we will briefly discuss some of these representations here.

We have seen that the Fourier transform of a Gaussian is still a Gaussian and, according to (4.1) and (4.2), it turns out that the spreads of these two Gaussians vary inversely as a function of α. More precisely, whereas the variance of $g_{\alpha,\beta}(t)$ is inversely proportional to α, the variance of its Fourier transform $G_{\alpha,\beta}(\omega)$ is proportional to the same α (the product of these two variances being kept constant). In terms of interpretation, this behavior (illustrated in Figure 5.1) can be simply phrased as follows:

> The larger the spread in one domain (time or frequency), the smaller the spread in the other one.

Remark. While Gaussians do play a special role, this behavior is very general and it can indeed be observed for any Fourier pair. In the context of physics, this has been underlined from the very beginning of quantum mechanics, especially by Werner Heisenberg, whose phenomenological considerations led to what is now referred to as his "uncertainty principle" [52]. Soon after, Hermann Weyl was amongst the first ones to give this phenomenon a purely Fourier mathematical perspective [53] and, in communication theory, the "time-frequency uncertainty" surfaced again when derived by Dennis Gabor in his 1946 seminal paper [29].

5.1   Variance

Given any square integrable signal $x(t) \in L^2(\mathbb{R})$ with finite energy $\|x\|_2^2$, it follows from the isometry of the Fourier transform (also known as Parseval's relation) that

$$\int_{-\infty}^{\infty} |x(t)|^2 \, dt = \int_{-\infty}^{\infty} |X(\omega)|^2 \, \frac{d\omega}{2\pi} = \|x\|_2^2, \tag{5.1}$$

and $|x(t)|^2|$ and $|X(\omega)|^2$ can legitimately be interpreted as energy densities (in time and frequency, respectively).

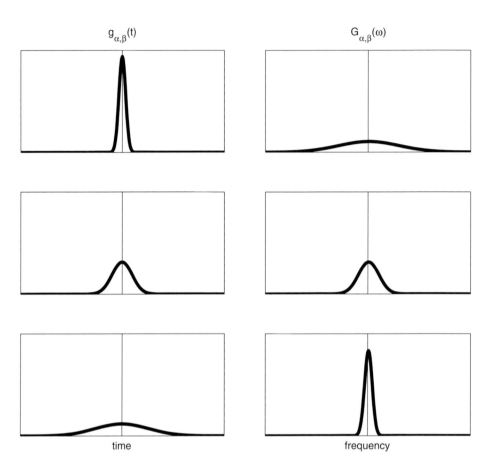

Figure 5.1 Gaussian waveforms and their Fourier transforms: shape-invariance and uncertainty. The left column displays Gaussians in the time domain, and the right one their Fourier transforms in the frequency domain, which are themselves Gaussians. The three rows correspond to three different values of α, evidencing the inverse variation of spreads in time and frequency.

Assuming for simplicity's sake that $\|x\|_2^2 = 1$, both densities do present the characteristics of a probability density function, and their spread can be measured in terms of variances. Without any loss of generality, it suffices to consider the case where both densities are centered around the origin, so that those variances Δt^2 and $\Delta \omega^2$ read

$$\Delta t^2 = \int_{-\infty}^{\infty} t^2 |x(t)|^2 \, dt \tag{5.2}$$

and

$$\Delta \omega^2 = \int_{-\infty}^{\infty} \omega^2 |X(\omega)|^2 \, \frac{d\omega}{2\pi}. \tag{5.3}$$

It then follows from a simple calculation involving the Cauchy-Schwarz inequality that

$$\Delta t \, \Delta \omega \geq \frac{1}{2}, \tag{5.4}$$

with the lower bound exactly attained by Gaussians of the form (4.1), with the parameterization $\beta = (2\alpha/\pi)^{1/4}$ to guarantee the unit norm. Of course, the same minimum uncertainty property holds true for the whole family (4.14) of time-frequency shifted versions of such waveforms, provided that the second-order moments (5.2) and (5.3) are evaluated with respect to the corresponding centers of the densities, i.e., t_0 and ω_0.

Gaussians, together with their time-frequency shifted versions, appear therefore as waveforms with minimum time-frequency uncertainty, making of them ideal candidates for being the "elementary" signals to be used as the most versatile building blocks for any complicated waveform, in the spirit of Gabor-like expansions.

> In Gabor's terminology [29], such Gaussian functions are referred to as "logons," which suggests that they are "elementary" in the sense that they carry a "quantum" of information that cannot be broken down further into smaller units.

Remark. The inequality (5.4) is established by assuming that there exists no coupling between time and frequency. Following Erwin Schrödinger [54], however, this assumption can be relaxed, resulting in the refined inequality

$$\Delta t \, \Delta\omega \geq \frac{1}{2} \, \sqrt{1 + \kappa^2}, \tag{5.5}$$

where κ is a "covariance" term that is expressed as:

$$\kappa = \int_{-\infty}^{\infty} t \, |x(t)|^2 \, \frac{d}{dt} \arg x(t) \, dt. \tag{5.6}$$

As explained in [22], this covariance can be thought of as an average, with respect to the energy density $|x(t)|^2$, of the product between time and the "instantaneous frequency" $\omega_x(t)$ defined as the phase derivative of the signal. If these two quantities are *independent*, one can expect that

$$\kappa = \overline{t \, \omega_x(t)}^{|x|^2} = \overline{t}^{|x|^2} \, \overline{\omega_x(t)}^{|x|^2}, \tag{5.7}$$

with the convention that $\overline{F}^{\rho}$ stands for the average of the function F with respect to the density ρ. If we note [22] that

$$\overline{\omega_x(t)}^{|x|^2} = \overline{\omega}^{|X|^2}, \tag{5.8}$$

the zero-mean assumption we made for the time and frequency densities leads directly from the vanishing of κ in (5.5) to the recovery of (5.4). However, when κ is nonzero, the lower bound is increased, and its maximum value is reached when the coupling between t and $\omega_x(t)$ is maximized. It follows from the Cauchy-Schwarz inequality that this happens in the colinear case corresponding to a quadratic phase. This means that minimizers of the Schrödinger inequality (5.5) are of the form

$$x(t) = b \exp\{a \, t^2\}, \tag{5.9}$$

with $a, b \in \mathbb{C}$ and the condition $\mathrm{Re}\{a\} < 0$. While such waveforms are referred to in physics terms as "squeezed states" – as opposed to the "coherent states" defined in (4.1) – in signal processing, they correspond to "linear chirps" with a Gaussian envelope.

5.2   Entropy

While the variance-based uncertainty amply justifies the use of Gaussians as "optimum" elementary signals, one can wonder whether their role would remain the same if some other measures of time and frequency spreading were considered. One such other measure that comes immediately to mind is *entropy*. In contrast to second-order property only, entropy is more general in that it relies on the full distribution, i.e., on all moments. From a statistical physics point of view, it is a measure of disorder: the more disordered a system, the more spread its probability density function and, hence, the greater its entropy. This follows from the simplest form of entropy, as initially defined by Ludwig Boltzmann in statistical physics and later made popular by Claude E. Shannon in information theory:

$$\mathcal{H}(p) = - \int_{-\infty}^{\infty} p(u) \log_2 p(u) \, du. \tag{5.10}$$

In continuation with the probabilistic interpretation of time and frequency energy densities, we can apply this definition to the time and frequency distributions ($|x(t)|^2$ and $|X(\omega)|^2$, respectively) of unit energy signals and, following Isidore I. Hirschman Jr [55] and William Beckner [56], we get the inequality:

$$\mathcal{H}(|x|^2) + \mathcal{H}(|X|^2/2\pi) \geq \log_2(\pi e). \tag{5.11}$$

Sharply localized distributions would correspond to small entropies but, again, what the inequality tells us is that there is a lower bound for the sum of the entropies attached to the energy densities in time and in frequency: they cannot both be made arbitrarily small. And, again, it can be proved that [55, 56]:

The entropic lower bound is attained for Gaussians.

In this way, the peculiarity of Gaussians is reinforced. More than that, the Shannon entropy defined in (5.10) enters a general class of entropies proposed by Alfréd Rényi as

$$\mathcal{H}_\alpha(p) = \frac{1}{1-\alpha} \log_2 \left(\int_{-\infty}^{\infty} p^\alpha(u) du \right), \alpha \in \mathbb{R}_+^* \backslash \{1\}, \tag{5.12}$$

with the property that

$$\mathcal{H}_1(p) = \lim_{\alpha \to 1} \mathcal{H}_\alpha(p) = \mathcal{H}(p). \tag{5.13}$$

Within this class too, it can be proved that the uncertainty inequality (5.11) can be generalized to

$$\mathcal{H}_\alpha(|x|^2) + \mathcal{H}_\alpha(|X|^2/2\pi) \geq \frac{\log_2 \alpha}{\alpha - 1} + \log_2 \pi \qquad (5.14)$$

for $\alpha \geq 1$, with Gaussians as minimizers.

5.3 Ubiquity and Interpretation

This chapter has provided a brief look at the question of uncertainty; more detail on this subject can be found in other works, e.g., in [57]. Further study of this subject is not within the scope of this book, but we will end here with the following comments.

1. First of all, uncertainty is *ubiquitous* and it shows up in many domains because it is essentially a by-product of Fourier analysis. It has been considered here for the time and frequency variables, but it applies equally as well to any Fourier pair. Although it was first formalized in the context of quantum mechanics, with position and momentum in place of time and frequency, it would apply similarly to other Fourier pairs such as space and spatial frequency.

2. The second comment is that, whereas different measures may be used for quantifying uncertainty, Gaussians happen to be minimizers for many of them. This is one of the reasons why Gaussians play a special role in time-frequency analysis and will be used thoroughly in the following chapters.

3. Finally, one can comment that the word "uncertainty" can be somehow misleading. Generally speaking, "uncertainty" has a nondeterministic flavor. As such, it may suggest that there would exist some "certainty" that would not be attainable because of either some intrinsic irreducibility (as in quantum mechanics) or of statistical fluctuations. This is indeed the case for estimation problems involving well-defined quantities such as time delay and Doppler shift in radar or sonar: in such situations, estimation is faced with a constraint that puts a (Cramér-Rao-type) limit to variance/covariance [58]. This corresponds to a setting that is actually statistical but which differs from the time-frequency setting that we previously considered, in which the second-order measures of spread for energy densities are deterministic and have no actual *variance* interpretation. An alternative interpretation in terms of *moment of inertia* would be better suited, with the term "uncertainty" replaced by "time-frequency inequality." Common usage has, however, chosen to adopt the more appealing term of "uncertainty." We have followed this usage and we will follow it henceforth, keeping in mind that, no matter what method "beyond Fourier" we may try to develop, the limit fixed by time-frequency inequalities will always apply in one form or another. In other words:

> There is no point in trying to "defeat Heisenberg" – not because there would exist some underlying reality with better localization properties, but simply because the question makes no sense as long as it is phrased in the Fourier framework.

6 From Time and Frequency to Time-Frequency

The "uncertainty" relationships that hold between the time and frequency representations of a signal underline how much the descriptions they offer are exclusive of each other. Both are facets of the same reality, but – just as with frontal and profile views in a portrait – the more we know about one, the less we know about the other. Acknowledging this fact suggests thinking in terms of projections and dimensionality. Rendering a 3D object on a 2D sheet of paper or on a photographic plate is a challenge and, rather than resorting to holographic techniques or the cubist style used by artists such as Picasso, it would prove more efficient not to reduce dimensionality and to make use, whenever possible, of sculpture (or 3D printing). In the time-frequency case, signals and spectra can be seen as 1D projections of some 2D "object." This is the rationale for switching from time *or* frequency to time *and* frequency, with both variables used jointly as is current practice in musical notation. Given this program and taking into account the constraints we have identified so far, how to write the musical score of a signal? The answer to this question is what this chapter proposes to address, not with the objective of making an exhaustive study of the options – since there are so many possible pathways to follow – but rather for the sake of stressing motivation and interpretation in the presentation of the main approaches that have been developed over the course of the past fifty or so years.

6.1 Correlation and Ambiguity

When introducing the time-frequency transform (4.16), we used the Gaussian waveform $g_{\alpha\beta}(t)$ exclusively as an analyzing function, but its form as a Short-Time Fourier Transform (STFT) is not restricted to this choice. In fact, it can be envisioned for any unit energy short-time window $h(t) \in \mathbb{R}$ as well, according to:

$$F_x^{(h)}(t, \omega) = \int_{-\infty}^{\infty} x(s) \, h(s - t) \, \exp\{-i\omega(s - t/2)\} \, ds, \tag{6.1}$$

with the corresponding spectrogram defined as

$$S_x^{(h)}(t, \omega) = \left| F_x^{(h)}(t, \omega) \right|^2. \tag{6.2}$$

We will therefore spend some time within this more general framework to discuss a number of notions that will later be applied to the Gaussian case.

While the Fourier transform maps a 1D signal in the time domain to a 1D spectrum in the frequency domain (and operates the other way for its inverse transform), the STFT maps the same 1D signal to a 2D function which is now defined on the time-frequency plane. However:

> If a STFT is a 2D function of time and frequency, not every 2D function of time and frequency is a STFT.

Indeed, the mapping defined in (6.1) induces some structure in valid STFTs. One of the most striking manifestations of this induced structure is certainly the *reproducing formula*, which states that

$$F_x^{(h)}(t', \omega') = \iint_{-\infty}^{\infty} K(t', \omega'; t, \omega) \, F_x^{(h)}(t, \omega) \, dt \frac{d\omega}{2\pi}, \tag{6.3}$$

with

$$K(t', \omega'; t, \omega) = F_h^{(h)}(t' - t, \omega' - \omega) \, \exp\{i(\omega t' - \omega' t)/2\} \tag{6.4}$$

the so-called *reproducing kernel* of the transform [7, 19]. The meaning of this formula is quite clear:

> The values of a STFT at two different locations (t, ω) and (t', ω') cannot be fixed arbitrarily: they are intimately coupled, one value depending on all the other ones in an integral way.

The strength of this dependence is measured by the kernel (6.4) which, up to a pure phase term, simply expresses as the STFT of the short-time window. Thanks to a simple change of variables, it is easy to rewrite $F_h^{(h)}(t, \omega)$ as $A_h(-\omega, t)$, with

$$A_h(\xi, \tau) = \int_{-\infty}^{\infty} h\left(s + \frac{\tau}{2}\right) h^*\left(s - \frac{\tau}{2}\right) \exp\{i\xi s\} \, ds \tag{6.5}$$

the so-called *ambiguity function*, in radar/sonar terminology [60, 61].

Remark. By construction, an ambiguity function can be expressed as

$$A_h(\xi, \tau) = \langle h, \mathbf{T}_{\xi,\tau} h \rangle, \tag{6.6}$$

where $\mathbf{T}_{\xi,\tau}$ stands for the time-frequency shift operator defined by

$$(\mathbf{T}_{\xi,\tau} h)(t) = h(t - \tau) \, \exp\{i\xi(t - \tau/2)\}. \tag{6.7}$$

The name "ambiguity" originates from the problem of jointly estimating the distance and the velocity of a moving "target." In an active system, this information is communicated by the differences detected in a known emitted signal's return echo. In a first approximation, distance can be converted into a delay τ corresponding to the duration of the round-trip for the emitted signal, and relative velocity translates as a

Doppler-shift ξ [60, 61]. Comparing the signal and its echo via a correlation can be shown to be optimal for detection (in terms of output signal-to-noise ratio) in idealized situations where the returning echo is embedded in white noise: this is the concept of "matched filtering" [62]. At the output of such a matched filter, the coherent "signal" part precisely expresses as (6.5)-(6.6) when exploring all possible Doppler shifts. If such a function were to be perfectly peaked, there would be only one value where it exceeds the threshold used for detection, and the location of the maximum would give the actual value (ξ, τ) of the pair delay/Doppler. We will argue in Chapter 7 that this is unfortunately not the case. For a given threshold, this leads to an infinite number of possible pairs delay/Doppler, hence the term *ambiguity*. This limitation in the joint estimation of delay and Doppler echoes the third comment about uncertainty at the end of the previous chapter.

Generalizing upon the classical notion of *correlation function* that amounts to measure the (linear) similarity between a waveform and its shifted version:

> The ambiguity function appears to be a 2D correlation function, in both time and frequency.

Its essential support in the time-frequency plane is thus controlled in time by the correlation "radius" of the waveform, and in frequency by the same quantity computed on its spectrum. In the specific case of the unit energy Gaussian $g_\alpha(t) = g_{\alpha,(2\alpha/\pi)^{1/4}}(t)$, we get, for instance,

$$A_{g_\alpha}(\xi, \tau) = \exp\left\{-\frac{1}{4}\left(\frac{\xi^2}{2\alpha} + 2\alpha\tau^2\right)\right\}. \qquad (6.8)$$

This provides evidence that, while the influence of the reproducing kernel extends all over the plane in theory, it is reduced to a limited domain in practice. This is in clear accordance with the interpretation of a STFT as a redundant, overcomplete representation that can yet be sampled according to the shape and extent of the reproducing kernel. In the Gaussian case, this is exactly the rationale for switching from a STFT to a Gaussian expansion.

The idea of ambiguity function and reproducing kernel is illustrated in Figure 6.1. For the sake of simplicity, we fix $\alpha = 1/2$ in this figure, thus ending up with the specialized Gaussian window $g(t) = g_{1/2}(t)$, which is, for an obvious reason, referred to as "circular" since (6.8) reduces in this case to

$$A_g(\xi, \tau) = \exp\left\{-\frac{1}{4}\left(\xi^2 + \tau^2\right)\right\}. \qquad (6.9)$$

When fixing $\xi = 0$, we recognize the ordinary (time) correlation function of $g(t)$, with a variance that is twice that of the waveform. For symmetry reasons, the same holds in the frequency domain when $\tau = 0$. Graphically and symbolically (this will be made more precise later), we can justify this result by considering the effective time-frequency domain of $g(t)$ as a disk whose radius is controlled by the common variance in time and frequency (2 in the circular case (6.9)).

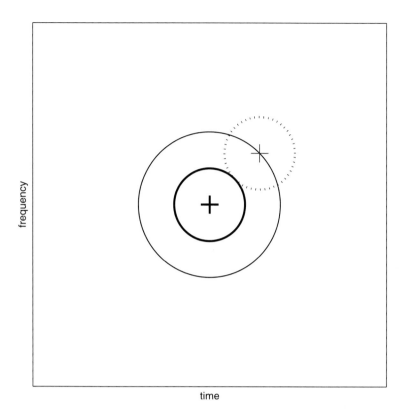

Figure 6.1 Reproducing kernel: a graphical interpretation. Whereas the thick circle bounds symbolically the essential time-frequency domain of an elementary Gaussian waveform, the support of the reproducing kernel of the STFT constructed with this waveform as a short-time window is itself bounded by the thin circle, which corresponds to the locus of all centers of time-frequency shifted circles (such as the dotted one) that are tangent to the thick circle, thus guaranteeing no overlap for the disks and, thus, no correlation.

Evaluating the time-frequency correlation amounts to measuring the overlap between this disk and any of its time-frequency shifted versions. Within this oversimplified interpretation, it is clear that the correlation vanishes as soon as the disks have no overlap.

It follows that the ambiguity function has a support, which is the disk whose circumference is given by the loci of all centers of the closest shifted disks that have no overlap with the reference one. The disk of the ambiguity function (and, hence, of the reproducing kernel) has therefore a radius twice that of $g(t)$.

6.2 Distribution and Wigner

Given that an ambiguity function is a kind of correlation function, we can advocate the very general duality principle that holds between correlations and distributions for getting access to a time-frequency distribution. Let us make this argument more clear. It is a well-known result of classical Fourier analysis that, for any deterministic signal $x(t) \in L^2(\mathbb{R})$, we have

$$\int_{-\infty}^{\infty} x\left(t + \frac{\tau}{2}\right) x^*\left(t - \frac{\tau}{2}\right) dt = \int_{-\infty}^{\infty} |X(\omega)|^2 \exp\{i\omega\tau\} \frac{d\omega}{2\pi}. \tag{6.10}$$

The left-hand side of expression (6.10) is a correlation function in the time domain, and the right-hand side equivalently expresses this quantity as the Fourier transform of an energy distribution in the frequency domain, the so-called *energy spectrum density*. This result is, however, more general, and if we rewrite the correlation function as $\langle x, T_\tau x\rangle$ it can be applied beyond this specific case. In particular, if we consider finite-power random signals and the associated inner product defined by $\langle x, y\rangle = \mathbb{E}\{x(t)y^*(t)\}$, where $\mathbb{E}$ stands for the expectation operator, $\langle x, T_\tau x\rangle$ stills appears as a correlation function, whose Fourier transform is now the *power spectrum density* (this is usually referred to as the "Wiener-Khintchine-Bochner theorem"). The framework is different but the interpretation is the same:

A distribution (in the sense of a density-like quantity, not of a generalized function) is the Fourier transform of a correlation, and vice-versa.

Moreover, while the distribution is expressed in its direct domain as a function of a variable which has an *absolute* meaning, the corresponding correlation expresses in the dual domain as a function of a variable that inherits a *relative* meaning. Indeed, whereas a spectrum density is a function of absolute frequencies, the correlation function depends on a time variable that has a status of lag, i.e., of a difference between absolute times. Conversely, if we had started from the temporal energy density $|x(t)|^2$ which is a function of absolute time, we would have ended up with a correlation function depending upon differences of frequencies (e.g., Doppler frequency shifts).

If we apply this Fourier duality principle to the ambiguity function (6.5) considered as a 2D time-frequency (lag-Doppler) correlation function, we readily get that

$$\iint_{-\infty}^{\infty} A_x(\xi, \tau) \exp\{i(\xi t + \omega\tau)\} d\tau \frac{d\xi}{2\pi} = W_x(t, \omega), \tag{6.11}$$

where

$$W_x(t, \omega) = \int_{-\infty}^{\infty} x\left(t + \frac{\tau}{2}\right) x^*\left(t - \frac{\tau}{2}\right) \exp\{-i\omega\tau\} d\tau. \tag{6.12}$$

This quantity, which can be considered as a time-frequency energy distribution since

$$\iint_{-\infty}^{\infty} W_x(t, \omega) dt \frac{d\omega}{2\pi} = \|x\|_2^2, \tag{6.13}$$

is referred to as the *Wigner* (or *Wigner-Ville*) *distribution* [30, 63].

It is worth noting that, besides the heuristic derivation outlined above, the Wigner distribution admits at least two closely related, yet different interpretations.

Interpretation 1 – Correlation function. The first interpretation amounts to see (6.12) as the Fourier transform of an *instantaneous* (or *local*) *correlation function.* This becomes particularly natural when switching from finite energy deterministic signals to finite power random processes and defining a time-dependent power spectrum according to $\mathbb{E}\{W_x(t, \omega)\}$. Under mild conditions [7], this can be expressed as

$$\mathbb{E}\{W_x(t, \omega)\} = \int_{-\infty}^{\infty} r_x\left(t + \frac{\tau}{2}, t - \frac{\tau}{2}\right) \exp\{-i\omega\tau\}\, d\tau, \tag{6.14}$$

with $r_x(u, v) = \mathbb{E}\{x(u)x^*(v)\}$.

Referring to Chapter 8 for a more specific discussion about stationarity, it then follows that a simplification occurs in the case where $x(t)$ happens to be *second-order stationary*, i.e., when there exists a function $\gamma_x(\tau)$ such that $r_x(u, v) = \gamma_x(u - v)$. Indeed, it is easy to check that the Wigner spectrum (6.14) boils down in this case to the time-independent quantity $\Gamma_x(\omega)$, which is the Fourier transform of the stationary correlation function $\gamma_x(\tau)$, i.e., exactly the power spectrum density. The Wigner spectrum appears therefore as a natural time-dependent extension of the stationary power spectrum.

Interpretation 2 – Characteristic function. The second interpretation goes back to the analogy between an energy density and a probability density. In this picture, the duality is now between probability densities and *characteristic functions*, and it suffices to first consider a characteristic function of time and frequency for getting a corresponding time-frequency distribution by a 2D Fourier transformation. This was the approach followed by Jean Ville in 1948 [30], leading him to propose the ambiguity function (6.5) as the simplest candidate for a characteristic function. In applying this duality to the ambiguity function (6.5), we immediately obtain the Wigner distribution (6.12) to which the name of Ville is sometimes attached.

We will not go into all the details of time-frequency distributions here (interested readers are referred to, e.g., [7] or [21] for a more comprehensive treatment), but it is worth mentioning that:

> Whatever the "distribution" interpretation (energy or probability), it is only an analogy that quickly finds its limits. In particular, in contrast with $|x(t)|^2$ or $|X(\omega)|^2$, which (for a correct normalization) can be considered as proper densities because they are non-negative, $W_x(t, \omega)$ is known to attain negative values for almost all signals.

Interestingly, one can remark that Gaussian waveforms are once more singled out in this context, since Robin L. Hudson proved in 1974 [64] that (up to an extra quadratic phase factor), they are the only ones to have a Wigner distribution that is non-negative everywhere. Wigner distributions of 1D Gaussians are indeed 2D Gaussians, as can be checked either from a simple direct calculation, or by Fourier inverting (6.8).

6.3   Spectrograms, Cohen, and the Like

The constructions of a time-frequency energy distribution in Section 6.2 are "natural" in some sense, but the resulting Wigner distribution is in no way the only solution. To be convinced of this fact, it is enough to be reminded that ordinary spectrograms, as defined in (6.2), are also admissible time-frequency energy distributions since, provided that the short-time window $h(t)$ is of unit energy, we have:

$$\iint_{-\infty}^{\infty} S_x^{(h)}(t, \omega) \, dt \frac{d\omega}{2\pi} = \|x\|_2^2. \tag{6.15}$$

In fact, there are infinitely many admissible time-frequency energy distributions, and whole classes of solutions can be derived from infinitely many covariance principles. Fortunately enough, prominent classes emerge on the basis of simple requirements. This is especially the case for the so-called *Cohen's class* $C_x(t, \omega; \varphi)$ that comprises all quadratic energy distributions that are covariant with respect to time-frequency shifts, i.e., such that

$$C_{\mathbf{T}_{\tau,\xi}x}(t, \omega; \varphi) = C_x(t - \tau, \omega - \xi; \varphi). \tag{6.16}$$

This class simply depends on a largely arbitrary parameterization function $\varphi(\xi, \tau)$ according to [7, 22]

$$C_x(t, \omega; \varphi) = \iint_{-\infty}^{\infty} \varphi(\xi, \tau) \, A_x(\xi, \tau) \, \exp\{i(\xi t + \omega \tau)\} \, d\tau \frac{d\xi}{2\pi}. \tag{6.17}$$

As compared to (6.11), this simply amounts to weighting the ambiguity function $A_x(\xi, \tau)$ by the parameterization function $\varphi(\xi, \tau)$ prior to the 2D Fourier transform. It thus follows that the Wigner distribution necessarily belongs to Cohen's class, with $\varphi(\xi, \tau) = 1$ as a parameterization function.

Spectrograms being quadratic and covariant to time-frequency shifts, they should also belong to this class. This is of course the case, and one simple proof (and characterization) goes as follows. One can first remark – and check by a direct calculation – that the Wigner distribution satisfies an isometry property which is referred to as *Moyal's formula* [7] and reads, for any two finite energy signals $x(t)$ and $y(t)$:

$$\iint_{-\infty}^{\infty} W_x(t, \omega) \, W_y(t, \omega) \, dt \frac{d\omega}{2\pi} = |\langle x, y \rangle|^2. \tag{6.18}$$

Given (6.1) and (6.2), it thus follows that $S_x^{(h)}(t, \omega) = |\langle x, \mathbf{T}_{t,\omega}h \rangle|^2$ and, hence, that

$$S_x^{(h)}(t, \omega) = \iint_{-\infty}^{\infty} W_x(\tau, \xi) \, W_{\mathbf{T}_{t,\omega}h}(\tau, \xi) \, d\tau \frac{d\xi}{2\pi}. \tag{6.19}$$

Making use of the covariance of the Wigner distribution with respect to time-frequency shifts, we readily get the alternative definition of a spectrogram as a smoothed Wigner distribution:

$$S_x^{(h)}(t, \omega) = \iint_{-\infty}^{\infty} W_x(\tau, \xi) \, W_h(\tau - t, \xi - \omega) \, d\tau \frac{d\xi}{2\pi}, \tag{6.20}$$

that will be of paramount importance in the following. Recognizing in the right-hand side a 2D correlation and using (6.11), the sought-after parameterization of the spectrogram within Cohen's class immediately follows as $\varphi(\xi, \tau) = A_h^*(\xi, \tau)$.

> Beyond the two considered cases of the Wigner distribution and of spectrograms, many other possibilities are offered for choosing specific parameterization functions whose structure reflects properties of the associated distribution.

Remark. The general form of Cohen's class as composed of "smoothed" Wigner distributions is intimately related to the chosen constraint of covariance with respect to time and frequency shifts. Other covariance requirements would of course end up with other classes of solutions, and such possibilities have been thoroughly explored in the literature (see, e.g., [21]). Without going into detail, it is useful to highlight the so-called *affine class* [65] which results from the covariance with respect to time shifts and dilations. What is remarkable is that such a (time-scale) class can still be organized around the Wigner distribution, all members of the class resulting in "affine smoothed" Wigner distributions. Copying the construction (6.2), the counterpart of spectrograms within the affine class is then given by *scalograms* (i.e., squared wavelet transforms).

Having mentioned the affine class, we will stop our study of these variations at this point and will only comment on the form (6.17) of Cohen's class. If we remember that the Wigner distribution results from a Fourier transformation upon the ambiguity function (which is considered as a correlation function), its extension to Cohen's class simply amounts to first applying a "taper" prior to Fourier transforming. This is reminiscent of an *estimation* procedure in classical spectrum analysis, such as the well-known technique of windowed correlograms [66]. Provided that $\varphi(\xi, \tau)$ is low-pass in the ambiguity domain, such a taper interpretation makes sense, resulting in a smoothing action on the Wigner distribution that tends to reduce fluctuations – though at the expense of a loss of resolution – with, therefore, a flavor of *bias-variance trade-off.*
We will come back later to the subject of fluctuations reduction, including a geometrical perspective, but at this point we can illustrate the loss of "resolution" when switching from the Wigner distribution to a spectrogram. This is illustrated in Figure 6.2, which gives an example of the smoothing relationship (6.20).

> It can furthermore be noticed that the very same relationship gives a more precise meaning to the symbolic representation of the reproducing kernel displayed in Figure 6.1.

Indeed, considering an ambiguity function as a "self-windowed" STFT, we can write:

$$A_g(\xi, \tau) = \left(\iint_{-\infty}^{\infty} W_g(\tau, \xi) \, W_g(\tau - t, \xi - \omega) \, d\tau \, \frac{d\xi}{2\pi} \right)^{1/2} \tag{6.21}$$

by noting that $A_g(\xi, \tau)$ is real-valued and everywhere non-negative. Since it is easy to show that $W_g(t, \omega) = 2 \exp\{-(t^2 + \omega^2)\}$, the disks of Figure 6.1 have to be understood as isocontours of Wigner distributions.

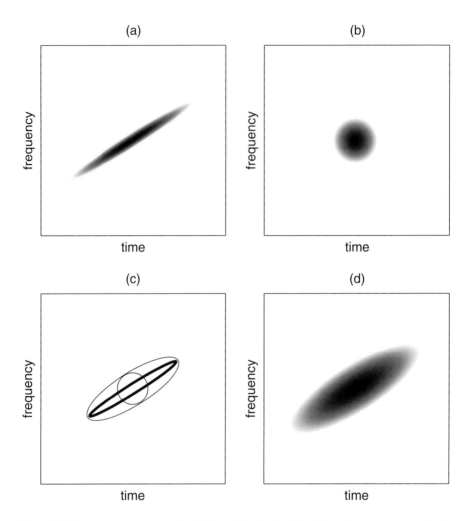

Figure 6.2 Spectrogram as a smoothed Wigner distribution. The Gaussian spectrogram of a linear chirp signal with a Gaussian amplitude modulation (d) results from the time-frequency smoothing of the corresponding Wigner distribution (a) by the Wigner distribution of an elementary Gaussian waveform (b). The symbolic representation of the corresponding domains is displayed in (c).

Before we proceed further, let us close this chapter with some (partly historical) remarks in the spirit of the "golden triangle" discussed in Chapter 1. Indeed, as for Fourier and wavelets, the Wigner distribution also originates from physics – more precisely, from quantum mechanics. The motivation for its introduction in 1932 (as a quasi-probability distribution function of position and momentum) was to permit conventional calculations in semi-classical situations [63]. Although this approach has some limitations, it led to a sustained interest in physics during the following decades, with mathematical facets (such as the theory of pseudo-differential operators) that proved to be of major importance in harmonic analysis (see, e.g., [67]). Initially, the Wigner distribution was no time-frequency distribution but – as it has been remarked when mentioning Weyl's derivation of uncertainty relations – what applies to a Fourier pair is applicable to any other one, *mutatis mutandis*. This is therefore the case with a function of position and momentum such as the original Wigner distribution, which can be turned into a function of time and frequency.

The first appearance of the Wigner distribution as a time-frequency distribution is due to Jean Ville [30], though with no reference to Wigner. This contribution, however, remained confidential – partly because it was originally written in French in *Câbles et Transmissions*, a "low-impact" journal, and partly because during that (pre-digital) time there existed no easy way to effectively compute it. When such possibilities developed in the 1960s, almost nobody remembered Ville's work, and it was only in 1980 that the idea of using a time-frequency Wigner distribution in signal processing resurfaced, with a series of papers by Theo A. C. M. Claasen and Wolfgang F. G. Mecklenbräuker [68]. This seminal work attracted considerable interest and made Wigner-based time-frequency analysis a "hot topic" in signal processing until the late 1990s [69]. Meanwhile, wavelets had appeared on the scene, and their popularity and efficiency overshadowed the Wigner distribution. Indeed, although wavelets offer more of a multiresolution ("time-scale") approach than a genuine time-frequency one, most newcomers interested in "going beyond Fourier" just took the wavelet shortcut, even if it was not necessarily the approach best suited for, say, instantaneous frequency estimation. In any event, the main reasons why the Wigner distribution did not become as standard as wavelets can be found in the "golden triangle": computing a Wigner distribution is cumbersome, no real fast algorithm exists, and extensions to higher dimensions are prohibitively complicated.

Although the Wigner distribution has not been adopted because of the broken symmetry of the "golden triangle," this should not be taken to mean that its study was (and still is) useless. It is of course a beautiful object that deserves study *per se* due to its many properties and interpretations in physics and mathematics, but this is not the only reason that it merits study. Indeed, a deep knowledge of its structure proved instrumental in deriving more efficient procedures (one can think of techniques to be discussed in the next chapters, such as, e.g., "reassignment," which partly regain the lost summit of the "golden triangle" regarding computation).

For many reasons, only some of which have been evoked here, the Wigner distribution is *central* in time-frequency analysis and it can be viewed as the forerunner of all time-frequency distributions. Maybe some of its derivatives have proved more successful, but, as with Newton's assertion that he was '*standing on the shoulders of giants*', forgetting about the Wigner distribution would be a mistake.

7 Uncertainty Revisited

As we have mentioned in previous chapters, regardless of the way a signal is described with respect to a Fourier pair of variables, it will never escape uncertainty. This holds true for all representations, leading to specific forms of inequalities that are explicitly stated for functions that use the time and frequency variables simultaneously. Some of those inequalities, which parallel what has been discussed in Chapter 5, are presented and discussed in this chapter.

7.1 L_2-Norm

We have seen that a Wigner distribution is an energy "distribution" in the sense that the result of its integration over the entire time-frequency plane equals the signal energy. Furthermore, because of Moyal's formula (6.18) and of Parseval's relation, we are assured that

$$\iint_{-\infty}^{\infty} |A_x(\xi,\tau)|^2 \, d\tau \frac{d\xi}{2\pi} = \iint_{-\infty}^{\infty} W_x^2(t,\omega) \, dt \frac{d\omega}{2\pi} = \|x\|_2^4. \tag{7.1}$$

Together with the magnitude inequalities $|A_x(\xi,\omega)| \le \|x\|_2^2$ and $|W_x(t,\omega)| \le 2\|x\|_2^2$ (which both follow from a direct application of the Cauchy-Schwarz inequality), this clearly means that:

> Both the (squared) ambiguity function and the Wigner distribution cannot concentrate their non-zero volume over a domain that would be arbitrarily small.

This is a joint form of time-frequency uncertainty that holds for any finite energy signal, and which can be given a somewhat more precise form as follows: given a time-frequency domain D of area $|D|$, if

$$\iint_D |A_x(\xi,\tau)|^2 \, d\tau \frac{d\xi}{2\pi} \ge (1-\epsilon) \|x\|_2^2 \tag{7.2}$$

for some $\epsilon > 0$, then we have necessarily $|D| \ge 1-\epsilon$. This was put forward by Karlheinz Gröchenig in 2001 [23], as a direct application of the Cauchy-Schwarz inequality.

7.2 L_p-Norms and Entropy

Energy-type inequalities are directly attached to an L_2-norm of the considered time-frequency quantities. More generally, L_p-norms can be considered in place of the L_2-norm as proposed by Eliott Lieb, who proved in 1990 [70] that:

$$
\begin{cases}
\|A_x\|_p \geq B_p \|x\|_2^2 & \text{for} \quad p < 2 \\
\|A_x\|_p \leq B_p \|x\|_2^2 & \text{for} \quad p > 2,
\end{cases}
\tag{7.3}
$$

where $B_p = (2/p)^{1/p}$. In all cases, inequalities are sharp, with the bound attained by Gaussians. This remarkable role of Gaussians, which both maximize and minimize the L_p-norm, depending on whether p is greater or smaller than 2, is reinforced by an entropic inequality which, for unit energy signals, reads:

$$
\mathcal{H}\left(|A_x|^2\right) = - \iint_{-\infty}^{\infty} |A_x(\xi, \tau)|^2 \, \log_2\left(|A_x(\xi, \tau)|^2\right) \, d\tau \frac{d\xi}{2\pi} \geq \log_2 e,
\tag{7.4}
$$

with equality when $x(t)$ is again a Gaussian.

Remark. In the case where $x(t)$ is real-valued and even, we have $W_x(t, \omega) = 2A_x(-2\omega, 2t)$, with the result that the inequalities and bounds that have been obtained for ambiguity functions apply *mutatis mutandis* to Wigner distributions.

Remark. For the sake of not multiplying notations, the joint Shannon entropy considered in (7.4) makes use of the same symbol $\mathcal{H}$ as in the previous univariate cases. We can proceed the same way for Rényi entropies of order α, with generalized inequalities that equally point towards the optimality of Gaussians when $\alpha \geq 1$; see [71] for details.

7.3   Concentration and Support

Coming back to the question of energy concentration over a given time-frequency domain, a companion approach is to consider the problem of finding the unit energy signal $x(t)$ that maximizes the quantity

$$
E_x(D) = \iint_{-\infty}^{\infty} G_D(t, \omega) \, W_x(t, \omega) \, dt \frac{d\omega}{2\pi},
\tag{7.5}
$$

with $G_D(t, \omega)$ some non-negative, "low-pass" time-frequency weighting function supported over a given "effective" domain D. This partial energy $E_x(D)$ can be shown to result from the action of a projection operator over D, whose kernel is related to the weighting function $G_D(t, \omega)$. From this perspective [7, 72]:

> Maximizing $E_x(D)$ can be turned into an eigenproblem, the solution of which is given by the eigenfunction with the largest eigenvalue.

Given the symmetries of the Wigner distribution, a natural choice is to restrict ourselves to "elliptical" domains D controlled by some time and frequency supports T and Ω thanks to a parameterization of the form

$$G_D(t, \omega) = G\left((t/T)^2 + (\omega/\Omega)^2\right), \tag{7.6}$$

with G some nonnegative, nonincreasing weighting function such that $G(0) = 1$. Doing so, it turns out that the problem depends on T and Ω only through the time-bandwidth product ΩT which, up to a factor of $\pi/4$, measures the area of the elliptical domain. Furthermore, and regardless of the chosen form for G, the eigenfunctions are given by Hermite functions, the first of which (of order $n = 0$) are simply the Gaussian function.

To be more specific, we can recall results obtained with some specific choices for the weighting function [7, 72].

Example 1. Let the weighting function be the indicator function $G(s) = \mathbf{1}_{[0,1/4]}(s)$, which amounts to *strictly* limiting the domain D to an ellipse with time and frequency axes of respective lengths T and Ω. Arranging the eigenvalues $\{\lambda_n(\Omega T); n = 0, \ldots\}$ (corresponding to Hermite functions as eigenfunctions) in decreasing order, it can be shown that, for any $n > 0$,

$$\lambda_n(\Omega T) < \lambda_0(\Omega T) = 1 - \exp\{-\Omega T/4\}. \tag{7.7}$$

> Once more, Gaussians stand out when they are being used for maximum concentration of signal energy in both time and frequency.

Example 2. Alternatively, let $G(s) = \exp\{-s^2/2\}$, which corresponds to an *effective* domain D whose support is controlled by the quantities T^2 and Ω^2, which can be interpreted as variances in time and in frequency, respectively. In this case, the eigenvalues are expressed as

$$\lambda_n(\Omega T) = \frac{\Omega T}{\Omega T + 1/2}\left(\frac{\Omega T - 1/2}{\Omega T + 1/2}\right)^n, \tag{7.8}$$

and we have again, for any $n > 0$, $\lambda_n(\Omega T) < \lambda_0(\Omega T)$, proving that Gaussians achieve the maximum energy concentration in the effective domain. Apart from this (expected) fact, it is worth noting that there exists a threshold value guaranteeing that all eigenvalues are nonnegative, namely $\Omega T = 1/2$. This is of course reminiscent of the variance-type uncertainty relation (5.4), whose lower bound is attained by Gaussians, and this is also in accordance with the nonnegativity of the spectrogram resulting from the smoothing (6.20) of the Wigner distribution, when restricted to Gaussian short-time windows [48]. In other words, smoothing a Wigner distribution with a 2D Gaussian does not necessarily result in a nonnegative distribution. This is only the case if the time-bandwidth product is such that $\Omega T \geq 1/2$. The lower bound of this condition is automatically attained when using the Wigner distribution of a Gaussian waveform as a smoothing function, as is the case for Gaussian spectrograms. From this point of view:

> Gaussian spectrograms can be viewed as minimally smoothed Wigner distributions that are always guaranteed to be nonnegative.

7.4   Variance

Although we will not explore all of the possible ways of characterizing the time-frequency uncertainty directly in the time-frequency plane, we will conclude this chapter with a variance-type approach that generalizes – in a simple joint way – what has been done previously, but separately, in time and frequency.

By considering time-frequency energy distributions $C_x(t, \omega; \varphi)$ that are members of Cohen's class (6.17), and elaborating upon the classical measures of spread defined in (5.2)–(5.3), we can introduce the quantity [7, 73]

$$\Delta^2(C_x) = \iint_{-\infty}^{\infty} \left(\frac{t^2}{T^2} + T^2\omega^2 \right) C_x(t, \omega; \varphi)\, dt\, \frac{d\omega}{2\pi}, \tag{7.9}$$

as a joint measure of the same nature, where T stands for a given reference time introduced for dimensionality purposes. Developing this expression, we get

$$\Delta^2(C_x) = \frac{\Delta t^2}{T^2} + T^2\Delta\omega^2 + c(\varphi), \tag{7.10}$$

where $c(\varphi)$ is a correction term that involves second-order partial derivatives of $\varphi(\xi, \tau)$ evaluated at the origin of the plane, and which thus vanishes in the Wigner case for which $\varphi(\xi, \tau) = 1$ [7].

Making use of the elementary property

$$\frac{\Delta t^2}{T^2} + T^2\Delta\omega^2 = \left(\frac{\Delta t}{T} - T\Delta\omega \right)^2 + 2\,\Delta t\,\Delta\omega \geq 2\,\Delta t\,\Delta\omega, \tag{7.11}$$

it readily follows from (5.4) that, in the Wigner case,

$$\Delta^2(W_x) \geq 1, \tag{7.12}$$

with equality in the Gaussian case.

If we now consider spectrograms attached to $\varphi(\xi, \tau) = A_h^*(\xi, \tau)$, it turns out that $c(A_h^*) = \Delta^2(W_h)$, thus leading to

$$\Delta^2(S_x^{(h)}) \geq 2, \tag{7.13}$$

with equality when both the signal and the short-time window are Gaussians. This adding up of the minimum spreads can be seen as a by-product of the relationship (6.20) that defines a spectrogram as a smoothed Wigner distribution.

Remark. In the case where the window $h(t)$ is "circular" and Gaussian (i.e., when it is chosen as $g(t) = \pi^{-1/4} \exp\{-t^2/2\}$), it follows from (6.9) that

$$\log\left(|A_g(\xi, \tau)|^2\right) = -\frac{1}{2}\left(\xi^2 + \tau^2\right).\tag{7.14}$$

Since $S_g^{(g)}(t, \omega) = |A_g(\omega, t)|^2$, we have in such a case an exact proportionality between the entropy and variance measures, namely:

$$\Delta^2(S_g^{(g)}) = \frac{2}{\log_2 e}\,\mathcal{H}(S_g^{(g)}),\tag{7.15}$$

thus recovering for both terms of this equation the lower bound 2 as a result of either (7.13) or (6.9).

7.5   Uncertainty and Time-Frequency Localization

Given the most classical and most constrained form of uncertainty inherited from (5.4), no perfect localization can be achieved in both time and frequency: the greater the concentration of signal energy in one domain (e.g., time), the smaller it is in the other (e.g., frequency), and vice versa.

When switching to time-frequency and considering either volume, support, or variance, the corresponding joint uncertainty relations forbid any *pointwise* localization of energy in both time and frequency. This can be easily understood when using the probabilistic interpretation of a time-frequency energy density, in which classical uncertainty measures (in time or frequency only) refer essentially to marginal distributions.

Given the flexibility offered by the existence of two degrees of freedom when working in the time-frequency plane, however, this does not rule out other, more general forms of localization along *trajectories*, seen as the limit of a "ribbon" whose width would go to zero while its extension would go to infinity, in such a way that the [finite] area is kept unchanged. This is illustrated symbolically in Figure 7.1, with a time-frequency localized linear chirp obtained as a limit of a *squeezed state* of the form (5.9) when $\text{Re}\{\alpha\} \to 0_-$ and $\text{Im}\{\alpha\} \neq 0$.

Indeed, the existence of a strictly positive lower bound in time-frequency uncertainty relations implies that any distribution necessarily extends over a time-frequency domain with some minimum, nonzero area. Given its area, however, the shape of this domain can be varied thanks to possible covariances in the plane. Within this picture, an initial "circular" logon can be either compressed or dilated in one variable to take the form of an ellipse, with a corresponding dilation or compression in the dual variable so as to keep the area unchanged. Such transformations can also be combined with rotations, eventually resulting in a linear localized structure in the limit case of an infinite stretching of the ellipse.

It is important to note that this illustration is not merely symbolic; it can also correspond to actual distributions. This is the case with the Wigner distribution, which takes on a 2D Gaussian form when applied to waveforms such as (5.9) (see Figure 6.2

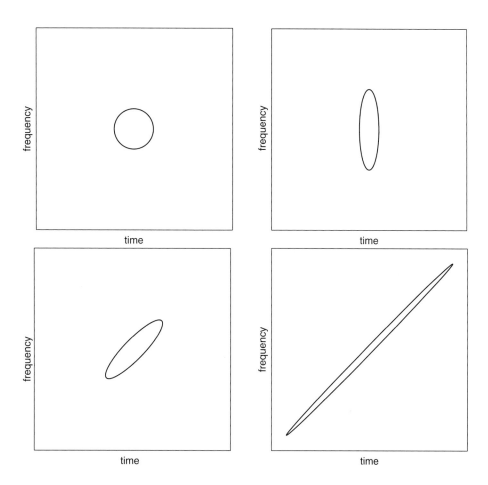

Figure 7.1 Uncertainty and localization. Without changing its nonzero area, a "circular" minimum uncertainty state with no pointwise localization (top left) can be successively squeezed (top right), rotated (left bottom), and stretched (bottom right) so as to end up with a linear chirp that is perfectly localized along the trajectory of its instantaneous frequency.

for a comparison with Figure 7.1), and which is known to localize perfectly along the straight line of these waveforms' instantaneous frequency in the limit case of a constant magnitude linear chirp of infinite duration [7, 22]:

$$x(t) = \exp\{ibt^2/2\} \Rightarrow W_x(t, \omega) = 2\pi\,\delta(\omega - bt). \tag{7.16}$$

Remark. The perfect localization of the Wigner distribution in the case of idealized linear chirps follows from the covariance of Cohen's class with respect to time-frequency shifts. This can be given a simple geometrical justification [74] that we will come back to at the end of Chapter 9, after having discussed the "interference geometry" of Wigner distributions.

8 On Stationarity

Before beginning the second part of this book (which will deal in a more detailed way with geometrical and statistical aspects of time-frequency distributions), it can be useful to discuss the concept of *stationarity*, which has already been advocated in several places of this book and that is routinely referred to in time-frequency analysis, but which is often used in a loose sense. Indeed, one can find in most introductions to time-frequency analysis the motivation of dealing with "nonstationary" signals, yet with different acceptions for nonstationarity. As a "non-property," nonstationarity can be thought of as the open area of infinite possibilities beyond stationarity, which in turn leads to the question: what is stationarity?

If we think in purely mathematical terms, the answer is clear and well-documented: stationarity is a concept applying to *random processes*, stating in a nutshell that distributional properties *do not depend upon some absolute time* [11, 43]. Stationarity can be defined for any statistical moments but, in signal processing, and in particular when dealing with Gaussian processes, *second-order* stationarity is of primary concern. In this case, stationarity of a random signal $x(t)$ is characterized by the two defining properties that (i) the mean value $\mathbb{E}\{x(t)\}$ is some constant m_x which does not depend on time, and (ii) that the covariance function $\mathbb{E}\{[x(t) - m_x][x(s) - m_x]^*\}$ only depends on the time *difference* $t - s$. The corresponding stationary covariance function is in Fourier duality with the (time-independent) power spectrum and it is clear that, when such a form of stationarity breaks down, time-frequency naturally enters the play so as to permit a time-dependent spectrum analysis.

While the mathematical framework is well-defined, it turns out to be far too rigid for practical applications (and interpretations), leading to twisted versions of the original concept that are of common use. First, stationarity is basically a *stochastic* notion, but it is standard to use the same terminology for *deterministic* signals whose spectral content is not changing with time. From this viewpoint, a sustained tone will be considered as stationary, while a chirp will not, connecting implicitly stationarity with *periodicity*. Second, the exact definition is quite demanding, since the translational invariance is a priori supposed to be verified for *all times* and *all lags*. Again, this assumption is often softened by introducing ideas of locality, with associated notions of *quasi-stationarity* or *local stationarity* [75]. Of course, this makes sense, but this also raises new difficulties regarding, e.g., the question of "testing for stationarity."

We will here argue that the question of (second-order) stationarity can gain from being re-addressed in time-frequency terms. Besides offering a framework that can encompass both deterministic and stochastic situations, this also permits us to reformulate stationarity as a *relative concept*, paving the way for testing stationarity in an operational sense [76].

8.1 Relative Stationarity

In his landmark book *The Fractal Geometry of Nature* [77], Benoit B. Mandelbrot gives an illuminating example for justifying why and how the concept of dimension may not be absolute, but makes the observer enter the picture. The example is as follows: '[...], a ball of 10 cm diameter made of a thick thread of 1 mm diameter possesses (in latent fashion) several distinct effective dimensions. To an observer placed far away, the ball appears as a zero-dimensional figure: a point. [...] As seen from a distance of 10 cm resolution, the ball of thread is a three-dimensional figure. At 10 mm, it is a mess of one-dimensional threads. At 0.1 mm, each thread becomes a column and the whole becomes a three-dimensional figure again. At 0.01 mm, each column dissolves into fibers, and the ball again becomes one-dimensional, and so on, with the dimension crossing over repeatedly from one value to another. When the ball is represented by a finite number of atom-like pinpoints, it becomes zero-dimensional again. [...] The notion that a numerical result should depend on the relation of object to observer is in the spirit of physics in this century and is even an exemplary illustration of it.'*

This key observation can be reproduced, *mutatis mutandis*, for illustrating that, from a practical point of view, the "stationarity" we usually (and, most often, implicitly) deal with in signal processing has to be similarly defined as a relative concept, depending on how we observe signals. In order to support this claim, let us consider the speech signal of Figure 8.1. In the first row, a 1 s long segment is displayed, with separate occurrences of three distinct vowels pronounced by a child (courtesy of Zoé Flandrin, 2003). At such a scale of observation, it will be commonly agreed upon that we face a "nonstationary" situation, loosely because of a time history that undergoes drastic changes. If we now focus on the middle vowel (middle row) so as to only display a 100 ms long segment, the quasi-periodicity of the waveform suggests a natural form of "stationarity" – and this is indeed what is assumed in linear predictive coding schemes that rely on the extraction of classical "stationary" attributes such as ARMA coefficients [141]. If we zoom in further on the signal down to an observation scale of 10 ms, however, we again face a "nonstationary" situation, with glottal pulses that are better acknowledged as individuals rather than members of a pulse train. What we learn from this example is that the perception we may have of the notion of "stationarity" is highly dependent on the observation scale, and that the very same signal may lead to intertwined interpretations that alternate between "stationarity" and "nonstationarity" when we travel across scales.

Remark. The interplay between a signal and its observation at a given scale is of course reminiscent of the *wavelet transform*, a possible companion tool for time-frequency analysis [19, 20]. Without elaborating on this transform, which we chose not to focus on

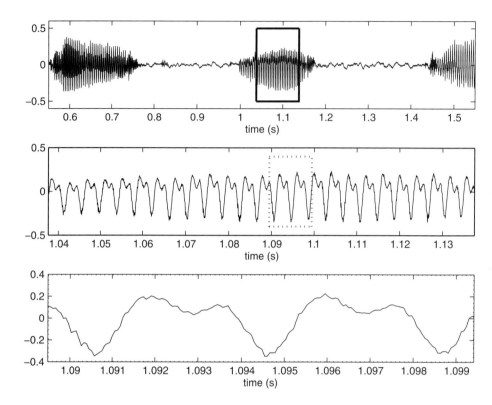

Figure 8.1 Relative stationarity. The top diagram corresponds to a speech segment observed over a time span of 1 s. The line box in the center of this diagram delineates a 100 ms long segment whose enlargement is displayed in the middle row. Similarly, the dotted line box in the middle diagram selects a 10 ms long segment whose enlargement is displayed in the bottom diagram.

in this book (see, e.g., [7] or [20] for links and connections), we can make the following comment. Given its definition

$$\mathcal{W}_x(t, a) = \frac{1}{\sqrt{a}} \int_{-\infty}^{\infty} x(s) \, \psi^* \left(\frac{s - t}{a} \right) \mathrm{d}s, \qquad (8.1)$$

the continuous wavelet transform $\mathcal{W}_x(t, a)$ is a function of time t and of a *scale* parameter $a > 0$, thanks to which the analyzing wavelet $\psi(t)$ is either dilated or contracted. To be admissible, this wavelet has to be essentially bandpass, leading to a time-frequency interpretation via the identification $\omega = \omega_0/a$, with ω_0 the characteristic central frequency of the wavelet filter at the reference scale $a = 1$. An example of the application of a continuous wavelet transform (based on a Morlet wavelet, i.e., a modulated logon) to a periodic pulse train is given in Figure 8.2. What is clearly revealed by this analysis is a continuum in the interpretation, from the finest scales where pulses are "seen" individually to the coarsest ones where the occurrence of many pulses within the time support of the wavelet results in a spectral line (and harmonics).

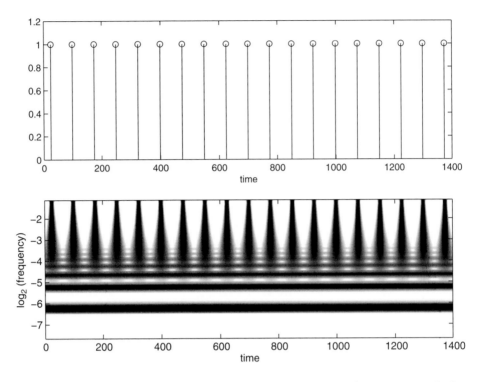

Figure 8.2 Wavelet analysis of a pulse train. The continuous wavelet transform (bottom row) of a periodic pulse train (top row) evidences a continuous transition from fine scales (high frequencies, where pulses are "seen" as individuals) to large scales (low frequencies, where periodicity ends up with spectral lines). Note that the frequency axis is displayed on a logarithmic scale.

Let us note that similar observations will be made for spectrograms in Chapter 16 (see Figure 16.13).

Coming back to the speech example of Figure 8.1, we can try to elucidate what makes us decide whether a signal, at a given observation scale, is stationary or not. Indeed, it is immediate to recognize that the decision follows from a comparison between *local* and *global* properties, within the chosen time span. The rationale can be precised further as follows.

We have previously mentioned that, for many reasons, a good candidate to the time-varying spectrum of a random signal is the Wigner spectrum, i.e., the expected Wigner distribution $\mathbf{W}_x(t, \omega) = \mathbb{E}\{W_x(t, \omega)\}$. Among those reasons is that if the process $x(t)$ happens to be stationary with power spectrum density $\Gamma_x(\omega)$, we then have for all times t the time-independent identification $\mathbf{W}_x(t, \omega) = \Gamma_x(\omega)$. In practice, we seldom have access to ensemble averaged spectra such as $\mathbf{W}_x(t, \omega)$ and we have to turn to estimators based on a single realization. In this respect, a spectrogram can be viewed as such an estimator [7].

As a smoothed Wigner distribution, the spectrogram computed on a realization of a stationary process (in the classical sense) is defined by a collection of frequency "slices" that are ideally all similar. This means that the local behavior attached to any time t should identify to the global behavior described by the marginal distribution:

$$\int_{-\infty}^{\infty} S_x^{(h)}(t, \omega)\, dt = \int_{-\infty}^{\infty} |H(\xi - \omega)|^2\, |X(\xi)|^2\, \frac{d\xi}{2\pi}. \tag{8.2}$$

Considering that the power spectrum density can be equivalently expressed as

$$\Gamma_x(\omega) = \lim_{T \to \infty} \mathbb{E}\left\{|X_T(\omega)|^2\right\}, \tag{8.3}$$

with

$$X_T(\omega) = \frac{1}{T} \int_{-T/2}^{T/2} x(t)\, \exp\{-i\omega t\}\, dt, \tag{8.4}$$

the right-hand side of (8.2) is just another way of thinking about the so-called Welch procedure for spectrum estimation [79].

8.2   Testing Stationarity

It follows from the previous considerations that a question such as "is this signal stationary?" does not have much absolute meaning. To make sense, it must be phrased relatively to an observation scale, specifying too the horizon over which the short-time analysis is performed. In other words:

> A realization of a signal observed over a given observation scale will be said to be stationary relative to this scale if its (spectrogram) local spectra undergo no time evolution.

Given an observation time span T_x, we thus have two scales at stake: the *global* one given by T_x itself, and a *local* one given by T_h, the (equivalent) time duration of the analysis window $h(t)$ of the STFT.

Remark. If we want to improve upon the estimation at the local scale, we could think of introducing a third subscale so as to apply the Welch procedure within the duration T_h. This would, of course, create a substantial increase in bias, and an alternative procedure is to make use a multitaper approach [76] (see Section 10.2 for a brief presentation of the multitaper approach).

Assuming that we are given the time span T_h and that we want to test for stationarity relative to this scale and a given window length T_h, one thing is clear: regardless of the estimation procedure and the chosen way of comparing between local and global spectra, we will never end up with a zero measure for the difference. We will always get some nonzero number, and the question is to decide whether this number is endowed with some significance. One possible answer can come from an *a contrario* perspective, based on the knowledge of what would be the natural fluctuations of the number chosen for testing in some comparable, but truly stationary situation, all things being equal regarding analysis. The chosen time-frequency framework offers a natural key to such a solution.

As an illustration, let us think of an observed signal of duration T, with an idealized *boxcar* spectrum density, flat over some bandwidth $B = [\omega_1, \omega_2]$ and zero elsewhere. This can arise from two very different situations. The first one results from the bandpass filtering of white noise over B, while the second one corresponds to a linear chirp, sweeping from ω_1 to ω_2 (or from ω_2 to ω_1) over T. The first situation can reasonably be thought of as stationary, while the second one cannot. The difference between the two cases is that, in the first case (filtered white noise), all Fourier modes associated with the frequencies within B are equally weighted but have no coherence in terms of their super-position in time, while the second case (linear chirp) is characterized by a strongly orga-nized structure within Fourier modes, so that while all underlying frequencies still have equal weight, they appear in an organized manner. Since all information about a signal is contained equivalently in its Fourier spectrum, and since both situations give rise to an identical spectrum *magnitude*, the difference must be found in the *phase* spectrum.

Indeed, in accordance with Fourier's shift theorem, the phase spectrum controls the way all Fourier modes are shifted with respect to each other, creating constructive interferences at some times and destructive interferences at other times. If the phase spectrum is random and uniformly distributed, no specific order emerges; in the case of a chirp, however, some organized structure leads to coherent evolutions, characterized essentially by trajectories in the time-frequency plane. This simple observation, which is illustrated in Figure 8.3, can be given a rigorous formulation as follows:

> Randomizing the phase spectrum while keeping the magnitude unchanged trans-forms a possibly nonstationary signal into a stationary one [76].

More specifically, this *stationarization* procedure amounts to (i) computing the Fourier transform of the observation, (ii) replacing the phase spectrum by a new random one that is uniformly distributed, and (iii) computing the inverse Fourier transform. Each randomization provides the user with *surrogate* data that is guaranteed to be stationary, while its magnitude spectrum is identical to that of the original data, by construction.

Remark. The concept of surrogate data was first introduced in nonlinear physics for sake of testing nonlinearity [80]. The question of stationarity was not addressed, but the reason for using a collection of suitably randomized versions of the observed data to characterize a hypothesis for possible rejection was the same.

To elaborate on the example contrasting a linear chirp and a bandpass filtered white noise shown on the previous page, Figure 8.4 illustrates the procedure of stationarization by surrogates. Modulating a linear chirp by a Gaussian envelope evidences a "nonsta-tionarity" in terms of the spectrogram signature and of its marginal distribution in time, both undergoing changes within the considered time span. When randomizing the phase

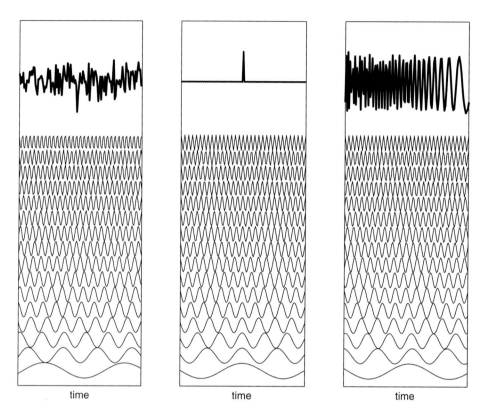

Figure 8.3 Phase spectrum and nonstationarities. In each diagram, the very same set of Fourier modes (thin lines) is considered, but with possibly different phase relationships. When no order is imposed (random phases, left diagram), the superposition of the modes (thick line at the top of the diagram) gives rise to random noise. In the middle diagram, phases are arranged so as to coincide exactly in the middle of the interval, resulting in an impulse. In the right diagram, phases have a local coherence which evolves as a function of time, resulting in a chirp.

spectrum while keeping the magnitude unchanged, we obtain a more erratic structure which tends to be identically distributed over time. The actual stationarization obtained in this way is clearly shown when a sufficient number of such surrogates are averaged together: as expected from the time-frequency interpretation of stationarity, the averaged spectrogram reproduces the magnitude spectrum for all times, with a constant marginal distribution in time as a by-product.

With this stationarization procedure at our disposal, it therefore becomes possible to create as many stationary surrogates as desired and to use them as a collection of realizations characterizing the null hypothesis of stationarity. This allows us to give a meaning to a "number" computed on the actual data, in comparison with the distribution of this number would be in the case of the data's being stationary.

A typical way of proceeding is to use a measure of the fluctuations in time of some "distance" between local spectra and the global spectrum obtained by averaging

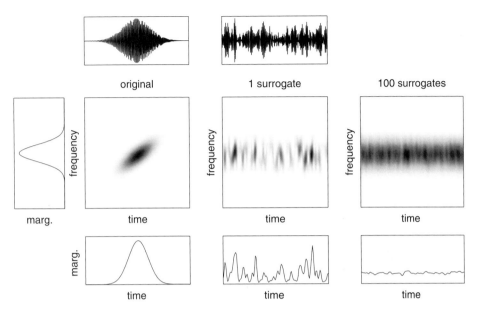

Figure 8.4 Time-frequency surrogates. The left group of subdiagrams displays a linear chirp with a Gaussian amplitude modulation in the time domain (top), its spectrogram (middle), and its marginals in time and frequency (bottom and left, respectively). The middle column plots the corresponding representations for one surrogate obtained by randomization of the phase spectrum (magnitude unchanged). The right column corresponds to an averaging over 100 such surrogates.

(or marginalization) as test statistics. More specifically, this can take the generic form [76]:

$$\lambda_x(T_h) = \frac{1}{T_x} \int_{T_x} \left(\Delta_x(t) - \langle \Delta_x(t) \rangle_{T_x} \right)^2 \, dt, \tag{8.5}$$

where $\Delta_x(t)$ stands for some dissimilarity measure between $S_x^{(h)}(t, \omega)$ and $\langle S_x^{(h)}(t, \omega) \rangle_{T_x}$, with the notation:

$$\langle \Phi(t) \rangle_T = \frac{1}{T} \int_T \Phi(t) \, dt. \tag{8.6}$$

Choosing a specific measure of dissimilarity depends on the type of nonstationarity we are interested in. From a time-frequency perspective, two main categories emerge naturally: frequency modulation (FM) and amplitude modulation (AM). An FM-type nonstationarity's signature is a spectrum whose shape changes over time, while, in the AM case, variations over time are expected to affect amplitude.

Based on the vast amount of information in the existing literature (see, e.g., [81]), this rough characterization suggests – analogous to what is done when comparing probability density functions – that we focus on the shape changes of the FM case by turning to a *Kullback-Leibler divergence* over a frequency domain Ω according to:

$$D_1 \left(S_x^{(h)}(t,.), \langle S_x^{(h)}(t,.)\rangle_{T_x} \right) = \int_\Omega \left(\hat{S}_x^{(h)}(t,\omega) - \langle \hat{S}_x^{(h)}(t,\omega)\rangle_{T_x} \right) \log \frac{\hat{S}_x^{(h)}(t,\omega)}{\langle \hat{S}_x^{(h)}(t,\omega)\rangle_{T_x}} \frac{d\omega}{2\pi},$$
(8.7)

where the notation "^" indicates a normalization to unit area.

As for the AM situation, a common and powerful measure is given by the *log-spectral deviation*, which takes on the form:

$$D_2 \left(S_x^{(h)}(t,.), \langle S_x^{(h)}(t,.)\rangle_{T_x} \right) = \int_\Omega \left| \log \frac{S_x^{(h)}(t,\omega)}{\langle S_x^{(h)}(t,\omega)\rangle_{T_x}} \right| \frac{d\omega}{2\pi}.$$
(8.8)

Of course, both AM and FM can occur simultaneously, leading to the combined measure

$$\Delta_x(t) = D_1 \left(S_x^{(h)}(t,.), \langle S_x^{(h)}(t,.)\rangle_{T_x} \right) \left[1 + \mu D_2 \left(S_x^{(h)}(t,.), \langle S_x^{(h)}(t,.)\rangle_{T_x} \right) \right],$$
(8.9)

where μ is a trade-off parameter that extensive simulations have suggested to take as unity [76].

Equipped with such a measure aimed at quantifying AM-FM-type deviations from a stationary situation, it becomes possible to estimate the probability density function of the test statistics (8.5) in the surrogate case by computing an empirical estimate based on the evaluation of $\lambda_{s_j}(t_h)$ for as many J surrogates $s_j(t)$ as desired. From our knowledge of this distribution, we can fix the threshold to be used for rejecting the null hypothesis of stationarity with a prescribed false alarm probability.

Constructed this way, the stationarity test is binary but it is expected that, the greater the nonstationarity, the larger the value of the observed test statistics as compared to the bulk of the surrogates distribution. This permits us to further quantify the nonstationary character of an observation by introducing an *index of nonstationarity* defined by:

$$\mathcal{I}_x(T_h) = \sqrt{\lambda_x(T_h) \left(\frac{1}{J} \sum_{j=1}^{J} \lambda_{s_j}(T_h) \right)^{-1}},$$
(8.10)

i.e., by measuring the ratio between the test statistics (8.5) attached to the actual observation $x(t)$ and the mean value of the same quantities computed on the J surrogates $\{s_j(t); j = 1,\ldots J\}$: the larger the former when compared with the latter, the greater the index.

As apparent in (8.5), the whole procedure still depends on the analysis scale given by the length T_h of the short-time window. By varying this quantity for a given observation span T_x (relative to which the stationary test makes sense), a typical *scale of nonstationarity* $\mathcal{E}_x$ can finally be defined according to:

$$\mathcal{E}_x = \frac{1}{T_x} \arg \max_{T_h} \{\mathcal{I}_x(T_h)\}.$$
(8.11)

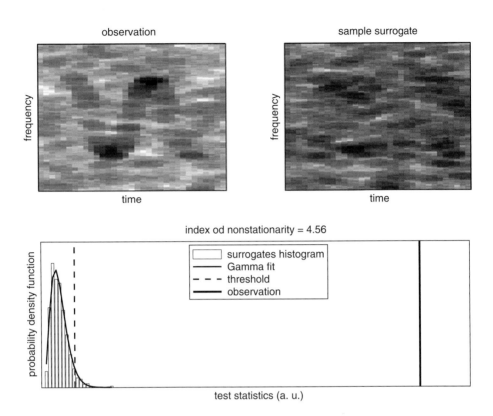

Figure 8.5 Testing for stationarity. The top left diagram displays the time-frequency distribution (multitaper spectrogram) of an AM-FM signal embedded in white Gaussian noise with an SNR of −6 dB. The top right diagram is similar, with the actual observation replaced by one realization of stationarized surrogate data. In both cases, $T_h/T_x = 0.2$, with T_h the short-time window length and T_x the total observation span relative to which stationarity is tested. The bottom diagram displays the value of the test statistics computed on the observation, to be compared with the threshold corresponding to a false alarm probability of 5%, derived from the distribution estimated from a set of 1,000 surrogates.

This achieves the goal of considering stationarity in an operational sense, offering a definition that is only relative (i.e., not absolute) by explicitly depending on an observation scale, and which includes the possibility of its testing thanks to the use of stationarized surrogates.

Whereas further details can be found, e.g., in [76], Figure 8.5 presents an example of the discussed method. It consists in the testing for stationarity over an AM-FM signal embedded in white Gaussian noise, with a ratio "analysis/observation" $T_h/T_x = 0.2$. The values obtained for the test statistics make clear that the observation is an outlier as compared to the threshold fixed by a given probability of false alarm (here 5%), which can be deduced from the empirical estimate of the probability density function attached to the stationary reference used as the null hypothesis of the test.

Remark. Interestingly, the histogram (here computed with 1,000 surrogates) turns out to be fairly well approximated by a Gamma distribution. This can be loosely justified from the form of (8.5) as a sum of squares resulting from a strong mixing. Assuming that such a model holds, the computational load of the test can be dramatically reduced, since it requires the estimation of only two parameters (namely, shape and scale), to determine what can be achieved in a maximum likelihood sense with $J \approx 50$ [76].

Remark. As an additional remark, the generation of stationary surrogates can be seen as the creation of a *learning set* composed of stationary references. This paves the way for making use of machine learning strategies (such as, e.g., one-class Support Vector Machines), offering a data-driven alternative to the model-based approaches constructed on distances [76].

Part II

Geometry and Statistics

9 Spectrogram Geometry 1

As we have seen repeatedly in the previous chapters, "logons," i.e., elementary Gaussian waveforms, play a very special role in time-frequency analysis. This is so because, in many respects, they correspond to the minimum "quantum" of information that can be accessed when considering time and frequency jointly. From now on, we will concentrate mostly on short-time analyses based on such Gaussian windows that, for both the sake of simplicity and reasons that will become clearer in the following chapters, will simply read

$$g(t) = \pi^{-1/4} \exp\{-t^2/2\} \tag{9.1}$$

so as to be of unit energy, and "circular" in the sense that

$$A_g(\xi, \tau) = \exp\{-(\xi^2 + \tau^2)/4\} \tag{9.2}$$

and

$$W_g(t, \omega) = 2 \exp\{-(t^2 + \omega^2)\}. \tag{9.3}$$

The corresponding *Gaussian spectrogram* $S_x^{(g)}(t, \omega)$, which results from squaring the representation $X_{1/2,\pi^{-1/4}}(t, \omega)$ when using formula (4.18), is referred to as the "Husimi distribution" in the physics literature [51]. Note that, in the interests of simplifying notation, we will from now on forget about the superscript "g" unless necessary.

Given those preliminary remarks, the purpose of this chapter is to explore the geometry that governs Gaussian spectrograms, using examples of progressively greater complexity.

9.1 One Logon

Let us start, therefore, with the simplest of the simple situations, the case of one isolated logon. There are many ways of computing the Gaussian spectrogram of such a waveform. This can be done either by a direct calculation from the definition (6.2), by using the smoothing relation (6.20) applied to (9.3), or by using (9.2) when simply remarking that $S_g(t, \omega) = |A_g(-\omega, t)|^2$. Whatever the method used, the result is given by

$$S_g(t, \omega) = \exp\{-(t^2 + \omega^2)/2\}. \tag{9.4}$$

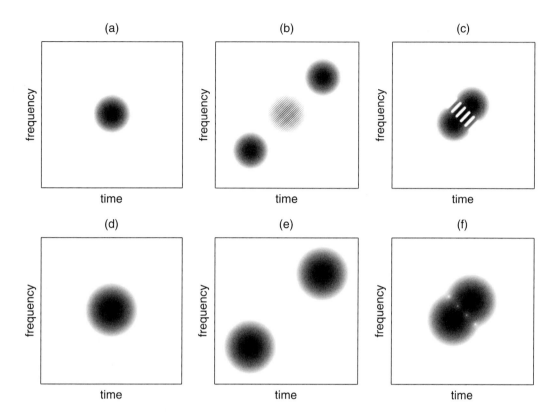

Figure 9.1 Wigner distributions (top row) and Gaussian spectrograms (bottom row) for one logon (left column) and two logons (middle and right columns). Comparing (a) and (d) (the one-logon case) illustrates the spreading due to the smoothing relationship shared by the Wigner distribution and the Gaussian spectrogram. The same applies to (b) and (e) as well as (c) and (f) (the two logons case) with the additional effect that oscillatory interference terms that exist in the Wigner distribution are smoothed out when the components are sufficiently far apart ((b) and (e)), while they create zeros and local maxima when they are close ((c) and (f)).

This is illustrated in Figure 9.1(d). There is not much to say about this distribution, except that it turns a 1D Gaussian into a 2D Gaussian and that, as expected, its effective time-frequency support is twice that of the Wigner distribution displayed in Figure 9.1(a).

9.2 Two Logons

A more interesting situation is encountered when switching from one logon to two, based on the following elementary observation:

> A spectrogram being structurally a quadratic transform according to (6.2), it necessarily obeys a quadratic superposition principle that is rooted in the well-known identity $(a + b)^2 = a^2 + b^2 + 2\,ab$.

In other words, when evaluating the squared sum of two quantities, we cannot simply add the two squared quantities; we must also include an extra term that accounts for possible interactions between the quantities at hand. In the specific spectrogram case we are interested in, this is expressed as

$$S_{ax+by}(t, \omega) = a^2 S_x(t, \omega) + b^2 S_y(t, \omega) + 2 ab \operatorname{Re} \left\{ F_x(t, \omega) F_y^*(t, \omega) \right\}. \qquad (9.5)$$

This shows that the extra term involves the product of the STFTs of the two considered signals. As such, it is essentially nonzero only in those time-frequency regions where the two STFTs overlap, in close connection with the idea of reproducing kernel that was discussed in Section 6.1. In the case where the two components $x(t)$ and $y(t)$ are sufficiently far apart in the time-frequency plane, they can be considered as "independent" in the sense that the spectrogram of their sum essentially reduces to the sum of the individual spectrograms. This is what can be observed in Figure 9.1(e). When the two components get closer as shown in Figure 9.1(f), however, this is no longer true; this case shows evidence of an interference structure that is characterized by the appearance of alternating zeros and local maxima.

> This is in fact the same situation as in classical interferometry, in which two shifted waves interfere and create what can be equivalently viewed as "fringes" in the time domain or "beating" in the frequency domain.

In order to better understand the mechanism that underlies this interaction phenomenon, it in helpful to turn to the Wigner distribution for which both Figures 9.1(b) and 9.1(c) show that such interactions do exist, whatever the spacing between the two components. To this end, we can model the two-component signal as $x(t) = g_+(t) + g_-(t)$, with

$$g_\pm(t) = g(t \mp \delta t/2) \exp\{\pm i(\delta\omega/2)t\}, \qquad (9.6)$$

where δt and $\delta\omega$ stand for the time and frequency spacings, respectively, between the two logons. A direct calculation ends up with the result

$$W_x(t, \omega) = W_g(t - \delta t/2, \omega - \delta\omega/2) + W_g(t + \delta t/2, \omega + \delta\omega/2) + I(t, \omega), \qquad (9.7)$$

in which the interference term $I(t, \omega)$ reads

$$I(t, \omega) = 2 W_g(t, \omega) \cos(t\,\delta\omega - \omega\,\delta t). \qquad (9.8)$$

As compared to the spectrogram, we see that, in the Wigner case, the interference term does not become any more diminished as the components move farther and farther apart: whatever the spacing, its maximum amplitude is unchanged (and even twice that of the individual components!). What is changed with spacing is the fine structure of this extra component which turns out to be oscillatory, hence with negative as well as positive contributions.

Although this example is simplified, it illustrates the general rules that control the *interference geometry* of the Wigner distribution, which can be summarized in a nutshell

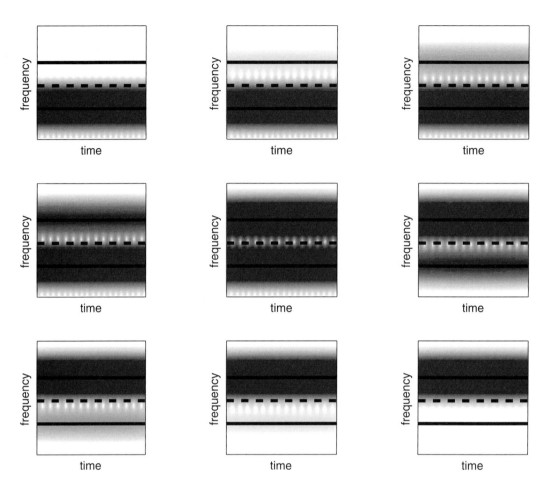

Figure 9.2 Gaussian spectrogram of a two-tone signal with unequal amplitudes. The figure displays, from top left to bottom right, a series of spectrograms when the amplitude ratio is varied, with the middle diagram showing an example of equal amplitudes. In each case, the superimposed horizontal lines indicate the frequencies of the two tones, and the dotted line which stands midway corresponds to the location of the interference term in the Wigner distribution. What this shows is that spectrogram zeros move from this location when the amplitude ratio differs from unity, in the direction of the component with the lowest amplitude.

as follows (interested readers are referred to [74] and [82] for a more comprehensive treatment of this interference issue):

> (i) interference terms are located *midway* between the interacting components, (ii) they *oscillate* proportionally to the time-frequency distance between the interacting components, (iii) the direction of oscillation is *orthogonal* to the straight line connecting the components.

Thanks to the smoothing relation (6.20), a spectrogram, while being everywhere non-negative by construction, can nevertheless attain zero values: this happens at precisely

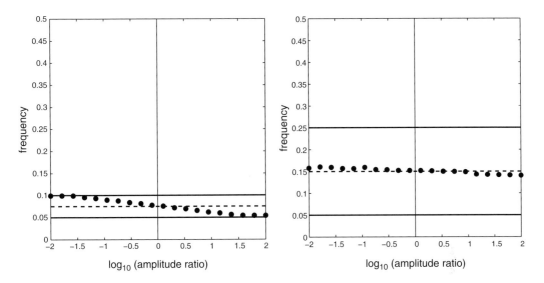

Figure 9.3 Frequency location of the Gaussian spectrogram zeros in the case of a two-tone signal with unequal amplitudes. While the horizontal lines indicate the frequencies of the two tones, and the dotted line which stands midway corresponds to the location of the interference term in the Wigner distribution, the dots refer to the frequency location of the spectrogram zeros, as a function of the amplitude ratio between the two tones. This provides evidence of a deviation with respect to the mid-frequency, the effect of which is more pronounced for closely spaced components (left) and tends to vanish when the components become sufficiently separated (right).

those points where positive and negative contributions of the interference term exactly compensate. In the case of equal amplitudes, this compensation occurs exactly midway between the two components, while an amplitude ratio different from unity makes the locus of the zeros move slightly in the direction of the lowest amplitude component. This is illustrated in Figure 9.2 with a two-tone example. Of course, as illustrated in Figure 9.3, this effect is more important for closely spaced components and tends to be negligible when the spacing is large enough, as compared to the reproducing kernel.

9.3 Many Logons and Voronoi

If we now consider a situation with many logons, the same arguments can be applied pairwise.

In the case of equal amplitudes, interactions between any two logons create an alignment of zeros along a segment that is perpendicular to the line joining the centers of the logons, as a natural consequence of the "mid-point geometry" discussed in the previous section. The global effect of this is the creation of organized edges, delimiting polygons that surround the centers of the logons – which essentially coincide with local maxima of the spectrogram. By construction, each time-frequency point within a given polygon is closest to the corresponding logon center than to any other logon center: this is the exact definition of a *Voronoi tessellation* [83].

> It is therefore remarkable that, in the case of many logons, zeros of the Gaussian spectrogram happen to be located on the edges of the Voronoi cells attached to *local maxima.*

Remark. Besides zeros and local maxima, which are located on the edges and at the centers, respectively, of such Voronoi cells, other points of interest can be considered, namely *saddle points.* It turns out that those specific points (which correspond simultaneously to a local maximum in some time-frequency direction and a local minimum in the orthogonal direction) also align along the edges, in between zeros.

We will come back to this geometric organization in further detail later, when considering a randomized instance of the many logons case as a simple model for complex white Gaussian noise. Meanwhile, this connection between local maxima, zeros, saddle points, and Voronoi cells is illustrated in Figure 9.4, with a variation encompassing unequal amplitudes.

Finally, as promised at the end of Chapter 7, we can have a new look at the time-frequency localization issue, from the interference geometry perspective that has been sketched in Section 9.2. Indeed, we can start from the observation that describing a signal in terms of subsignals, or "components," presents a part of arbitrariness. For instance, in the case of an infinite duration chirp extending over the whole real line, we can equally consider it as composed of two "half-chirps"; one half extends over the line at negative times, and the other half extends over at positive times. In general, it is possible to cut the chirp into as many discrete pieces as desired, provided they are non-overlapping and that their union covers the entire line. Within this picture, any two chirp pieces interact pairwise, creating an interference term located midway between their time-frequency locations, and the final Wigner distribution results from the combination of all such interactions, for all possible pairs. Therefore, if we consider a specific time-frequency point while repeating *ad infinitum* the procedure of dividing the chirp into smaller and smaller parts, the resulting interference at this point is made of all possible interactions between time-frequency "points" that are *symmetric* with respect to it, in accordance with feature (i) of the interference geometry principle (see Section 9.2). As a consequence of the companion features (ii) and (iii), the overall interferences combination results in the summation of infinitely many oscillations in the same direction, with all possible frequencies: this is precisely the Fourier definition of a Dirac's "δ-function." The reasoning has been made for one specific time-frequency point, but it applies equally well to any point along the instantaneous frequency line, thus justifying the perfect localization of the Wigner distribution in the idealized case of a unimodular, infinite duration linear chirp – as in (7.16).

Remark. Reversing the perspective, the specificity of linear chirps for guaranteeing a perfect localization of the Wigner distribution can be viewed as stemming from the geometrical property that a straight line is the only curve that is the locus of all of its

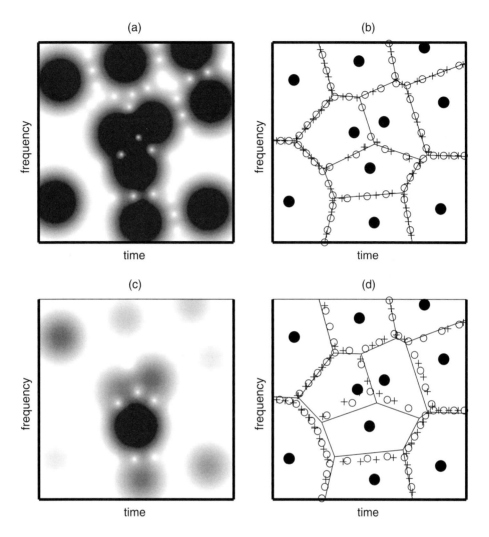

Figure 9.4 Gaussian spectrograms (left column) and Voronoi cells (right column). In the case of many logons with equal amplitudes (top row), zeros (circles) and saddle points (crosses) of the Gaussian spectrogram happen to be located on the edges of the Voronoi cells (lines) attached to the centers of the logons (dots) which also coincide with local maxima. This situation is similar in the case of unequal amplitudes (bottom row), though slightly modified, as explained in Figures 9.2 and 9.3.

mid-points. This can be generalized as well to some nonlinear chirps by using distributions that obey suitably modified covariance requirements and therefore obey alternative "midpoint" constructions [7, 84].

This way of considering localization as a by-product of interference sheds some new light on the role played by cross-terms. It is abundantly advocated in the literature that the existence of cross-terms – and their proliferation when the number of interacting components increases – severely impairs the readability of Wigner distributions. This is definitely true, but it has to be stressed that it is the very same mechanism that permits localization for linear chirps. In this respect, both a desirable property (localization) and a problematic one (cross-terms) go fundamentally together, the modification of the latter being achievable only at the expense of the former.

Classic spectrograms, as smoothed Wigner distributions, eliminate most of the oscillatory cross-terms while spreading potentially localized chirp components. This geometric form of trade-off naturally raises the question of whether the situation could be improved, i.e., of sharpening a spectrogram as far as its coherent, chirplike components are concerned, while minimizing the impact of cross-terms in between such incoherent, distinct components. This is precisely the topic of the next chapter.

10 Sharpening Spectrograms

As we have seen in (6.20), spectrograms result from the smoothing of the Wigner distribution of the signal by that of the short-time window. The advantage of this operation is twofold. First, this guarantees *nonnegativity*, a property that simplifies the interpretation of a spectrogram as an energy "density," in contrast with the Wigner distribution which locally attains negative values for almost all signals. Second, this nonnegativity also corresponds to the *smoothing out* of the oscillatory interference terms of the Wigner distribution that occur pairwise between any two components, be they close or not, and thus heavily hamper its readability. Smoothing does come with a drawback, however, since it *smears* components that are highly localized in the Wigner distribution.

The most extreme situation of localization is that of *linear chirps* whose simplest form is $c(t) = \exp\{i\gamma t^2/2\}$, with $\gamma \in \mathbb{R}$ the "chirp rate." This waveform can be thought of as the basic Gaussian (4.1) with a purely imaginary parameter $\alpha = -i\gamma/2$, and it can be generalized to include any time and/or frequency shift. In this idealized case, it is easy to check that $W_c(t, \omega) = \delta(\omega - \gamma t)$. As mentioned when discussing Figure 7.1, the Wigner distribution therefore happens to be *perfectly* localized on the "instantaneous frequency" line $\omega_x(t) = \gamma t$.

> In some sense, and whatever the chirp rate, the Wigner distribution behaves for one single linear chirp as the Fourier transform does for one single pure tone.

Remark. The perfect localization for linear chirps can be obtained by considering only pure tones and noting that, beyond time and frequency shifts, the Wigner distribution is also covariant to *rotations* in the time-frequency plane. In this respect, a chirp corresponds to the rotation of a tone with an angle $\theta = \tan^{-1}(\gamma)$. In the limiting case where $\theta \to \pi/2$, this approach even permits us to include the case of pure impulses, for which the Wigner signature is $\delta(t)$. Of course, all of these operations must be carefully considered and interpreted as limiting cases that cannot be actually achieved, since the corresponding idealized waveforms (tones, chirps, impulses) are neither integrable nor square integrable. Just as for the Fourier transform, a rigorous approach would amount to defining all time-frequency "distributions" as actual distributions, in the sense of generalized functions.

Based on these introductory observations, choosing between a spectrogram and a Wigner distribution is a matter of balancing the trade-off between localization of actual components and existence of "fake" interference terms.

> The Holy Grail of time-frequency analysis would therefore be to find some distribution which would be both nonnegative and as free of interference terms as a spectrogram, while being as localized as the Wigner distribution.

This quest for sharpened spectrogram-like distributions has led to a number of approximate solutions. Three of them will be outlined in the rest of this chapter.

10.1   Reassignment

The first approach starts again from the smoothing relationship (6.20). What this relation tells us is that *one number* – namely the value of the spectrogram at a given time-frequency point (t, ω) – summarizes (in a weighted way) the information contained by *the whole distribution* of values of the Wigner distribution within the effective domain $D(t, \omega)$ of the smoothing function surrounding the point of interest.

What it also tells us is that this number is simply assigned to the evaluation point (t, ω), whatever the distribution of $W_x(t, \omega)$ over $D(t, \omega)$. This makes sense when this distribution does present some symmetry around (t, ω) but, in the most general case, the evaluation point (t, ω), is not necessarily the best representative of the distribution.

> To illustrate this situation using a mechanical analogy, it is as though we assigned the total mass of an object to a point at the *geometric* center of the domain over which this mass is distributed, even though it is well known that this point must be located at the object's *center of mass*, or centroid.

Going back to the time-frequency interpretation, if we consider the idealized case of a linear chirp, we have seen that the Wigner distribution is perfectly localized along the instantaneous frequency line $\omega_x(t)$. The smearing of the corresponding spectrogram is therefore easy to understand: a nonzero contribution to the spectrogram will be observed in some evaluation point (t, ω) as long as the instantaneous frequency line will intersect the domain $D(t, \omega)$ centered around this point. In all cases, the centroid $(\hat{t}(t, \omega), \hat{\omega}(t, \omega))$ of $W_x(t, \omega)$ over $D(t, \omega)$ necessarily belongs to $\omega_x(t)$, but it is generally different from (t, ω), except in the very specific case where the evaluation point actually lies on $\omega(t)$.

The rationale of the so-called reassignment technique [85–88] is therefore quite natural:

> Reassignment amounts to simply moving the spectrogram value from the evaluation point to the local centroid by *reassigning* $S_x(t, \omega)$ from (t, ω) to $(\hat{t}(t, \omega), \hat{\omega}(t, \omega))$.

This results in a reassigned spectrogram $\hat{S}_x(t, \omega)$, which reads:

$$\hat{S}_x(t, \omega) = \iint_{-\infty}^{\infty} S_x(\tau, \xi) \, \delta\left(t - \hat{t}(\tau, \xi), \omega - \hat{\omega}(\tau, \xi)\right) \, d\tau \frac{d\xi}{2\pi}. \tag{10.1}$$

For this to be done, we need to know $\hat{t}(t, \omega)$ and $\hat{\omega}(t, \omega)$. As centroids of W_x weighted by W_h over D, they are expressed as:

$$\hat{t}(t, \omega) = \frac{1}{S_x(t, \omega)} \iint_{-\infty}^{\infty} \tau \, W_x(\tau, \xi) \, W_h(\tau - t, \xi - \omega) \, d\tau \frac{d\xi}{2\pi} \tag{10.2}$$

and

$$\hat{\omega}(t, \omega) = \frac{1}{S_x(t, \omega)} \iint_{-\infty}^{\infty} \xi \, W_x(\tau, \xi) \, W_h(\tau - t, \xi - \omega) \, d\tau \frac{d\xi}{2\pi}. \tag{10.3}$$

In practice, they will not have to be computed this way since it can be shown [87] that they can be expressed with the much simpler forms:

$$\hat{t}(t, \omega) = t + \text{Re}\left\{ \frac{F_x^{(\mathcal{T}h)}(t, \omega)}{F_x(t, \omega)} \right\} \tag{10.4}$$

and

$$\hat{\omega}(t, \omega) = \omega - \text{Im}\left\{ \frac{F_x^{(\mathcal{D}h)}(t, \omega)}{F_x(t, \omega)} \right\}, \tag{10.5}$$

with $(\mathcal{T}h)(t) = th(t)$ and $(\mathcal{D}h)(t) = dh(t)/dt$.

As compared to a standard spectrogram that requires the computation of one STFT with a short-time window $h(t)$, a reassigned spectrogram needs two additional STFTs involving two short-time windows that are simply derived from $h(t)$ (and even only one in the case where $h(t) = g(t)$, since $(\mathcal{T}g)(t)$ and $(\mathcal{D}g)(t)$ are then proportional).

We mentioned that the smoothing that underlies a spectrogram explains the smearing of localized components, and how reassignment counterbalances this effect and actually sharpens the distribution. The very same smoothing results in a reduction of oscillating interference terms between separated components, with an efficiency that is increased in proportion to the time-frequency spacing between the components, since the more widely spaced the components, the faster the oscillations, and therefore the easier it is to remove them by the action of a given smoothing function.

These findings are summarized in Figures 10.1 and 10.2.

Whereas it makes sense that reassignment recovers the perfect localization of the Wigner distribution in the limit case of a unimodular linear chirp of infinite duration, one can wonder what happens for an elementary logon. At first glance, the result is surprising since a direct calculation shows that

$$\hat{S}_g(t, \omega) = 4 \exp\{-2(t^2 + \omega^2)\}, \tag{10.6}$$

with the associated time-frequency uncertainty measure:

$$\Delta^2(\hat{S}_g) = \frac{1}{2}. \tag{10.7}$$

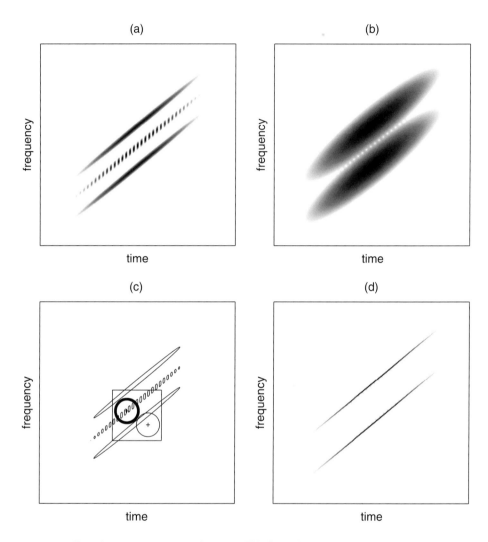

Figure 10.1 Gaussian spectrogram reassignment. This figure illustrates the efficiency of reassignment in the case of a signal made of two (amplitude modulated) linear chirps that are time-frequency disjointed while marginally overlapping in both time and frequency. The Wigner distribution is plotted in (a), the Gaussian spectrogram in (b), and the associated reassigned spectrogram in (d). In the lower left diagram (c), the circles superimposed on the simplified contour plot of the Wigner distribution represent selected examples of smoothing domains centered on interference terms (thick circle) and located in the vicinity of one of the two chirps (thin circle). The square delineates a domain that is magnified in Figure 10.2 for further explanations.

When we compare (10.6) with (9.4) and (9.3) on the one hand, and (10.7) with (7.13) and (7.12) on the other, this means that the reassigned spectrogram is not only more sharply localized than the spectrogram, but also more sharply localized than the Wigner distribution, suggesting that the uncertainty lower bound would be defeated. This is, however (and fortunately), not the case. In other words:

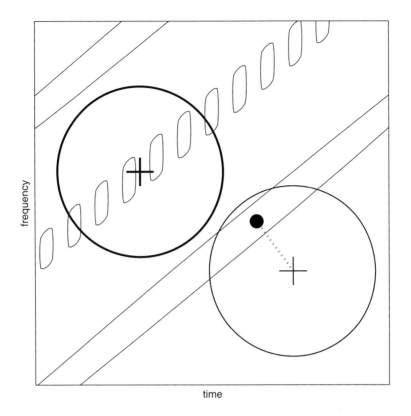

Figure 10.2 Gaussian spectrogram reassignment (close-up view). This figure is the enlargement of the part of Figure 10.1(c) that is contained within the square. It illustrates the role of the Gaussian smoothing function, whose support is essentially contained within a circle, depending on where the center (black cross) is located. In the case of the thick circle, the center lies in the middle of Wigner interference terms whose oscillations are smoothed out in the spectrogram. In the case of the thin circle, the center is off the chirp, yet sufficiently close to "see" it and contribute a nonzero value in the spectrogram. Reassignment counterbalances the smearing by moving this value to the centroid (dot) of the Wigner distribution within the considered domain.

> A sharp localization does not mean that Heisenberg is defeated: reassignment is not a "super-resolution" method.

Indeed, if this were true, if reassignment were acting as some "super-resolution" method, it would allow for the separation of closely spaced components beyond the Rayleigh limit that is fixed for any Fourier-based method, such as the spectrogram. As illustrated in Figure 10.3, we see that this does not happen: when components get close enough (in this case, midway between closely spaced linear chirps), an interference structure pops up.

From a more fundamental point of view, it must be understood that reassignment is a process that involves two ingredients. On the one hand, we have the reassigned distribution that can be seen as the final outcome of the process, and on the other hand

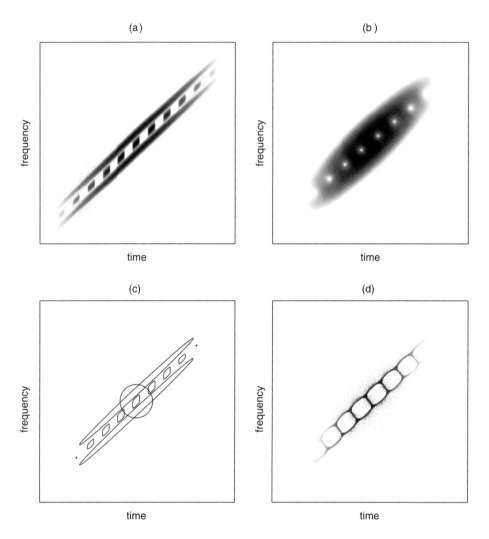

Figure 10.3 Reassignment sharpening does not mean super-resolution. This figure illustrates the resolution limitation of reassignment in the case of two (amplitude modulated) linear chirps that are closely spaced. The Wigner distribution is displayed in (a), the spectrogram in (b), and its reassigned version in (d). In the lower left diagram (c), the circle superimposed on the simplified contour plot of the Wigner distribution represents the smoothing domain centered on a time-frequency point midway between the two chirps. This domain also corresponds to a time-frequency zone where interference terms occur in the reassigned spectrogram (d).

we also have to consider the domain that consists of all those points that are reassigned. As for a Fourier description which, to be complete, needs to add a phase component to the only magnitude spectrum, the magnitude of a reassigned spectrogram must be complemented by a *vector field* that plays the role of a phase. We will come back to this point later when discussing the reassignment vector field and its relation to the idea of a *basin of attraction*. Meanwhile, it is worth noting that this phase interpretation is more than an analogy, since the very same reassignment operators that are given by (10.4) and (10.5) can be equivalently expressed as

$$\hat{t}(t, \omega) = \frac{t}{2} - \frac{\partial \Phi_x}{\partial \omega}(t, \omega) \qquad (10.8)$$

and

$$\hat{\omega}(t, \omega) = \frac{\omega}{2} + \frac{\partial \Phi_x}{\partial t}(t, \omega), \qquad (10.9)$$

where $\Phi_x(t, \omega)$ stands for the phase of the STFT $F_x(t, \omega)$.

Remark. At this point, we can make a comment about the "nonlinear" ways by which scientific ideas may evolve. Indeed, the idea of reassignment was first advanced by Kunihiko Kodera, Roger Gendrin and Claude de Villedary in the mid-1970s [85, 86]. At that time, they referred to it as a "modified moving window method," with an original formulation that was explicitly based on STFT phase relationships such as (10.8) and (10.9). This idea languished for many years for a variety of reasons: its proponents were outsiders to the signal processing community; their seminal paper was first published in a single, isolated journal; but the main reason was that the idea was simply ahead of its time, and almost nobody in the field was interested in the problem it addressed; this was later compounded by the fact that when the field began to take an interest in this problem, the existence of this idea had been forgotten. The idea surfaced again in the mid-1990s [87], with a new interpretation related to the smoothing of the Wigner distribution, thanks to advances made during the previous decade. This revisiting of the method offered new perspectives in at least two respects. First, the smoothing interpretation within Cohen's class allowed for extensions beyond the only spectrogram (and even to more general classes of distributions, e.g., time-scale distributions extending wavelet transforms). Second, it provided an alternative and efficient way of computing the reassignement operators, switching from the "differential" formulation (10.8)–(10.9) to the "integral" one given by (10.4)–(10.5). This resulted in a more versatile method that found applications in various areas, e.g., in acoustics and audio [89], and in any case beyond the purely geophysics domain from which it originated. In retrospect, it turns out that reassignment has many virtues as compared to Wigner-like distributions, and we can say that, nowadays, the latter are rarely used *per se*. Nevertheless, their detailed study has by no means been a waste of time: it has been a detour which has proved instrumental in the comprehension and building of tools that are ultimately better to use. Later on, with "synchrosqueezing" and "Empirical Mode Decompositions," we will see other examples of such a situation where the development of a side technique favors the revival of an older, forgotten idea.

10.2   Multitaper Reassignment

Up to now, we have focused on idealized situations of noise-free signals, with the objective of sharpening a spectrogram so as to be as localized as possible along time-frequency trajectories attached to instantaneous frequency laws. Ignoring noise is of course a limiting assumption, and taking it into account does prove interesting for at least two reasons. The first one is that noise is expected to affect the reassignment of localized components, thus calling for techniques aimed at reducing possible noise-induced disturbances. The second one is that noise may be of interest for its own sake, giving the problem an aspect of (nonstationary) spectrum estimation.

The way conventional spectrograms behave when analyzing (complex white Gaussian) noise will be explored in greater detail in Chapters 13–15, but we can present a few key facts here to give an idea of what is to come. By definition, white noise has a theoretical spectrum which is flat and, because it is also a stationary process, its "time-frequency spectrum" is expected to be flat as well, in both directions (time and frequency). This may happen to be true from an ensemble average perspective when using spectrograms, but, as is the case for the ordinary power spectrum, the situation is dramatically different when the analysis is restricted to one single realization. In this case, the time-frequency surface exhibits a high level of variability, alternating bumps and troughs in a random (though constrained) manner, and reassignment reinforces this variability by sharpening the bumps – as it would do for any local concentration of energy.

In classical spectrum estimation, it is well known that a crude estimator based on the only periodogram (i.e., the squared magnitude of the Fourier transform of the observed realization) is not consistent, with an estimation variance that is at the order of the squared quantity to be estimated. Reducing variance can only come from averaging "independent" estimations, with the constraint that those estimations cannot be obtained from different realizations but have to be inferred from the single one which is at our disposal. The classic technique – which goes back to Peter D. Welch in 1967 [90] – amounts to averaging short-time periodograms, each of them being computed on one of adjacent segments within the observation span. The rationale is that, if the correlation of the data is "microscopic," such segments will be almost uncorrelated (and, hence, independent in the Gaussian case). They can therefore serve as substitutes to unavailable extra realizations, but there is a price to be paid in terms of *bias* for the expected variance reduction. Indeed, if we are given an observation of total duration T, targeting to divide the estimation variance by a factor of, say, N implies chopping the data into N segments. Each of those segments then has a duration divided by N as compared to the total duration, degrading the frequency resolution by the same amount.

Remark. The Welch technique, which consists of averaging short-time periodograms, can be given a time-frequency interpretation [79]. If we think of the spectrogram of a signal $x(t)$ as a concatenated collection of squared local Fourier transforms with window $h(t)$ (in other words, of short-time periodograms), averaging them is equivalent to computing the marginal distribution of the spectrogram, up to a normalization constant. This marginal distribution is known to be a smoothed version of the global periodogram $|X(\omega)|^2$, namely:

$$\int_{-\infty}^{\infty} S_x^{(h)}(t, \omega)\, dt = \int_{-\infty}^{\infty} |X(\xi)|^2\, |H(\xi - \omega)|^2\, \frac{d\xi}{2\pi}, \tag{10.10}$$

thanks to which the transfer function $|H(\omega)|^2$ of the (low-pass) window acts as a regularizer for the fluctuations of the periodogram. One further point of interest of this interpretation is that it goes beyond spectrograms and applies in particular to *scalograms* (i.e., squared wavelet transforms), supporting their preferential use for a spectral-like analysis of scaling processes with power-law spectra [79].

In 1982, in order to overcome the bias-variance trade-off of the Welch procedure, David J. Thomson [66, 91] suggested an alternative technique that takes full advantage of the whole duration of the observation so as not to increase bias, while reducing variance thanks to an averaging based on a different kind of periodogram. The key point is to relate the decorrelation of the data-generated surrogate observations to a projection onto a set of orthonormal functions. From this point of view, the rationale of Thomson's approach is to replace adjacent, non-overlapping indicator functions with another system in which each member extends over the whole observation time span.

In contrast with short-time periodograms, which are classically based on *low-pass* windows that are most often nonnegative, the *multitaper* approach proposed by Thomson makes use of a family of windows that may oscillate. In the original formulation, this family was chosen to be that of the so-called *Prolate Spheroidal Wave Functions* [66, 91] because of their property of being maximally localized over a given frequency band, while being strictly limited in time.

The idea of multitapering can naturally be extended to a nonstationary setting, in which spectrograms based on a set of orthonormal functions (used as windows) are averaged. In this setting, it proves more advantageous to replace Prolate Spheroidal Wave Functions, which somehow dissymmetrize time and frequency, by functions that guarantee a maximum energy concentration in a symmetric way. As it has been argued in Chapter 7, Hermite functions $\{h_k(t); k = 0, 1, \ldots\}$ can play this role, in a natural generalization of the Gaussian window which happens to be the first of them. The multitaper spectrogram follows as being the average:

$$\bar{S}_x^{(K)}(t, \omega) = \frac{1}{K} \sum_{k=0}^{K-1} S_x^{(h_k)}(t, \omega). \tag{10.11}$$

One can give a simple justification of the efficiency of multitapering for a purpose of spectrum estimation as follows: As we have said previously, the Gaussian-based spectrogram of white Gaussian noise results in a time-frequency surface made of bumps, with energy patches located around local maxima. More generally, if we consider spectrograms computed with Hermite functions $h_k(t)$ as windows, it follows from the orthonormality of the system that $\langle h_k, h_{k'} \rangle = 0$ whenever $k \neq k'$, and a direct calculation shows that, in this case and for the very same realization $x(t)$:

$$\iint_{-\infty}^{\infty} F_x^{(h_k)}(t, \omega) F_x^{(h_{k'})*}(t, \omega) \, dt \, \frac{d\omega}{2\pi} = 0. \tag{10.12}$$

This suggests that the oscillations of $F_x^{(h_k)}(t, \omega)$ and $F_x^{(h_{k'})}(t, \omega)$ must compensate each other and that the local maxima of the corresponding spectrograms $S_x^{(h_k)}(t, \omega) = |F_x^{(h_k)}(t, \omega)|^2$ and $S_x^{(h_{k'})}(t, \omega) = |F_x^{(h_{k'})}(t, \omega)|^2$ must differ. This is indeed what is observed,

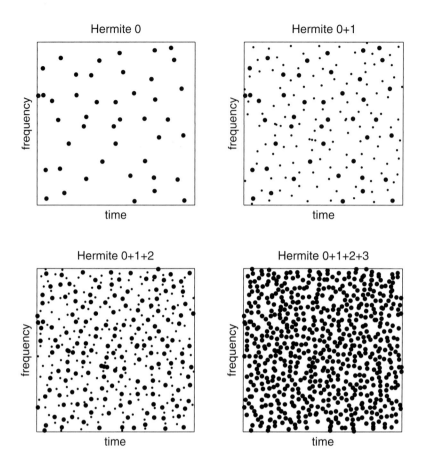

Figure 10.4 Multitaper rationale. Top left: local maxima of spectrogram of white Gaussian noise with a Gaussian (Hermite function of order 0) as a window. Top right: big dots reproduce the maxima of the top left diagram, and new maxima obtained with the Hermite function of order 1 are superimposed as small dots. Bottom left: big dots reproduce the maxima of the top right diagram, and new maxima obtained with the Hermite function of order 2 are superimposed as small dots. Bottom right: all maxima with Hermite functions of orders 0 to 3.

as illustrated in Figure 10.4, which shows how local maxima of the spectrograms are intertwined when using successive Hermite functions as short-time windows. We therefore obtain a coverage of the time-frequency plane that is more and more dense when adding up more and more Hermite spectrograms, resulting in an average spectrogram surface that becomes flatter and flatter, as would be the case when averaging conventional spectrograms computed on independent realizations together.

Figure 10.5 presents an example of the application of this technique to a synthetic noisy signal consisting of a nonlinear chirp embedded in a nonstationary noise obtained by filtering a time-limited segment of white Gaussian noise with a time-varying band-pass filter, whose central frequency is a linear function of time (the domain where most of the energy of the noise is supposed to be concentrated is indicated on the different

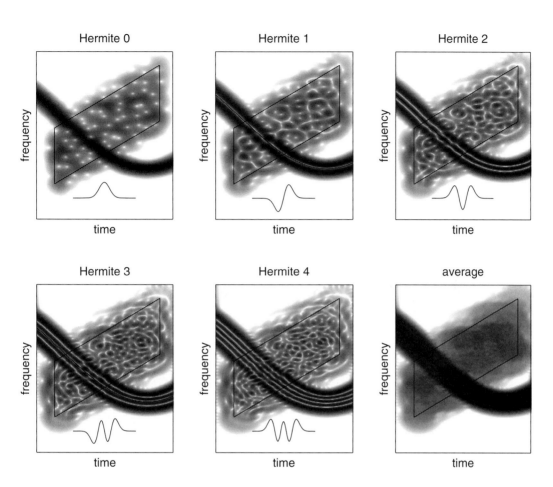

Figure 10.5 Multitaper spectrogram. Hermite spectrograms of increasing orders and their average, in the case of a nonlinear chirp embedded in a nonstationary noise whose effective time-frequency support is indictated by a thin, trapezoidal contour. For each diagram, the Hermite window is superimposed (at an arbitrary position).

diagrams by a thin, trapezoidal contour). Whereas the estimated spectrum is expected to be flat within the noise domain, it is clear that each individual spectrogram obtained using a Hermite function as a window exhibits fluctuations. As expected, the different local maxima occur at different locations in the plane and they become more and more numerous as the order of the Hermite function gets larger. The overall result is that the average of the different spectrograms tends to rub out the fluctuations and to provide a flatter estimated spectrum, though at the expense of a spreading of its time-frequency support. The same spreading effect is observed for the chirp component.

In order to improve upon this situation, the same strategy can be applied to reassigned spectrograms [92], as illustrated in Figure 10.6. We note in this case the spiky nature of the crude Gaussian spectrogram estimate of the noise, which goes along with a very sharp localization of the chirp along its instantaneous frequency. Here, the advantage

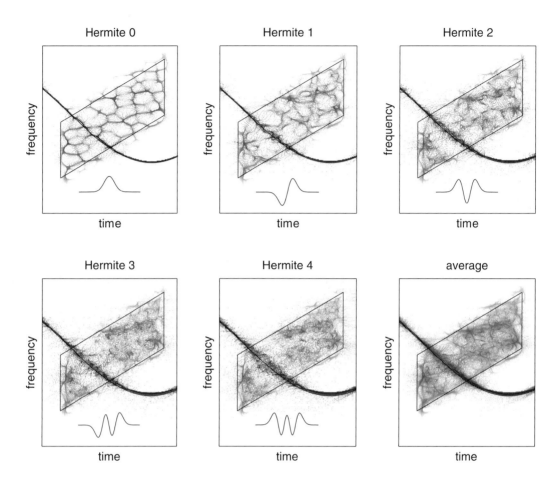

Figure 10.6 Multitaper reassigned spectrogram. Hermite reassigned spectrograms of increasing orders and their average, in the case of a nonlinear chirp embedded in a nonstationary noise whose effective time-frequency support is indicated by a thin, trapezoidal contour. For each diagram, the Hermite window is superimposed (at an arbitrary position).

of averaging reassigned spectrograms is twofold. First, it permits a regularization of the time-varying spectrum, while preserving its time-frequency support. Second, theory tells us that, in the locally linear approximation of the instantaneous frequency (at the scale of the short-time window), reassignment is almost perfect whatever the window. This is what is observed, with a localization which remains almost the same for all Hermite functions, in contrast with the spectrogram case.

10.3   Synchrosqueezing

Reassignment (and, by extension, multitaper reassignment) has many virtues but it is primarily a technique aimed at enhancing the localization and the readability of a time-frequency distribution. It is therefore essentially an *analysis* method which does present one main drawback: it is not invertible. A variation on reassignment is possible, however, allowing us to overcome this limitation. It is referred to as *synchrosqueezing* and can take on different forms. Initially formulated by Ingrid Daubechies and Stéphane Maes in a wavelet framework in the mid-1990s [93], it has more recently surfaced again on a STFT basis (see, e.g., [94]), and we will briefly discuss this case only.

> In a nutshell, synchrosqueezing operates a displacement of computed values in the time-frequency plane as reassignment does, but it differs from the latter in two respects. First, the reassignment operation only concerns the frequency direction. Second, it applies to the complex-valued STFT.

The frequency "reassignment" operator is therefore the same as the one previously considered, while the *synchrosqueezed* STFT $\tilde{F}_x(t, \omega)$ simply reads

$$\tilde{F}_x(t, \omega) = \int_{-\infty}^{\infty} F_x(t, \xi)\,\delta\,(t, \omega - \hat{\omega}(t, \xi))\,\frac{d\xi}{2\pi}, \qquad (10.13)$$

with $\hat{\omega}(t, \omega)$ as in (10.5) or (10.9).

Thanks to the displacement of STFT values towards the frequency centroid, the modified transform $\tilde{F}_x(t, \omega)$ is sharpened as compared to $F_x(t, \omega)$, hence the name "squeezing." Moreover, this sharpening is operated "vertically" at each time instant, hence the term "synchro."

Since synchrosqueezing simply amounts to reorganize (linearly) the distribution of STFT values at a given time, we have necessarily

$$\int_{-\infty}^{\infty} \tilde{F}_x(t, \omega)\,\frac{d\omega}{2\pi} = \int_{-\infty}^{\infty} F_x(t, \omega)\,\frac{d\omega}{2\pi}. \qquad (10.14)$$

This is particularly interesting from a reconstruction point of view since it is easy to check that – with the definition (6.1) we have chosen for the STFT – the analyzed signal can be exactly recovered according to the 1D formula

$$x(t) = \frac{1}{g(0)} \int_{-\infty}^{\infty} F_x(t, \omega)\frac{d\omega}{2\pi}, \qquad (10.15)$$

which is indeed a much simpler expression than the standard 2D inversion which – following from (4.16) and (6.7) – reads:

$$x(t) = \iint_{-\infty}^{\infty} F_x(\tau, \xi)\,(\mathbf{T}_{\tau\xi}g)(t)\,d\tau\frac{d\xi}{2\pi}. \qquad (10.16)$$

If we consider a "mode" with a reasonably slowly-varying instantaneous frequency, the STFT values to be synchrosqueezed are essentially contained within a ribbon surrounding this instantaneous frequency. Rather than making use of the whole frequency axis as in (10.15), one can therefore restrict the integration domain to some interval which measures the local width of the ribbon. This is of special interest in the case of multicomponent signals in which the different modes are sufficiently spaced apart so that the corresponding ribbons can be considered as disjoint.

> In such a situation, synchrosqueezing offers the twofold advantage of an enhanced representation (sharp localization along the instantaneous frequency) and a disentanglement of the composite signal in its different components (with reconstruction).

Of course, there is a price to be paid for this reconstruction capability, which is directly coupled with the fact that synchrosqueezing operates a "vertical" reassignment only, along the frequency direction. This is reasonably effective in the case of slowly-varying instantaneous frequencies, but runs into limitations whenever the variations turn out to be fast. An example is shown in Figure 10.7, in the very peculiar case of a Hermite function whose time-frequency structure corresponds – as an image of a harmonic oscillator – to circles in the plane. Whereas the Wigner distribution displays a complicated interference pattern, the spectrogram takes on the typical form of a "doughnut"-like, thickened circle. As expected (and as explained in greater detail in [95]), the reassigned spectrogram is sharply localized on the circle, with an equal concentration regardless of the local orientation of the time-frequency trajectory. This is due to the two-dimensional displacement, which tends to favor time reassignment in the vicinity of a local time-frequency structure which is "vertical" (and frequency reassignment when the structure is "horizontal"), with a balanced continuum in between. This flexibility, which is inherited from the rotation invariance of the Wigner distribution, does not hold for synchrosqueezing, which restricts reassignment to the only frequency direction. As a result, a sharp localization, comparable to that obtained by reassignment, is observed on the upper and lower parts of the circle where the "horizontal" instantaneous frequencies have least variation. In contrast, the localization is progressively decreased when approaching the extreme left and right parts of the circle, where the instantaneous frequencies tend to merge as locally "vertical" trajectories.

10.4   Sparsity

Besides reassignment and synchrosqueezing, which are post-processing methods, there is one more way of obtaining a spectrogram-like distribution with high localization, provided we make some reasonable assumptions about the time-frequency structure of the analyzed signal.

If we contemplate the examples used so far, and if we consider more generally simple, coherently structured signals (think of the chirps in Chapter 2, of speech, music and the like), a striking feature is the following:

> Even if a spectrogram has low time-frequency concentration, its nonzero (or non-negligible) values occupy only a small fraction of the available area of the time-frequency plane. In other words, the representation is *sparse*.

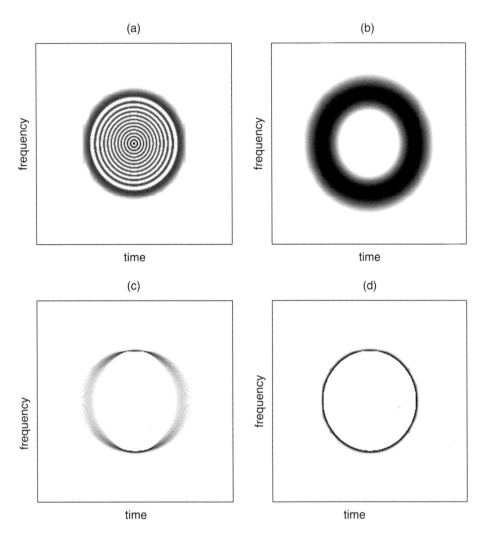

Figure 10.7 Synchrosqueezing versus reassignment. The time-frequency "trajectory" of a Hermite function (here of order 25) is a circle in the plane. In the Wigner distribution (a), this circle delineates a disk whose inner domain is filled with a complicated, oscillating interference pattern. Those oscillations are smoothed out in the Gaussian spectrogram (b), resulting in a "doughnut"-like structure corresponding to a thickened circle, which is dramatically sharpened after reassignment (d). Synchrosqueezing (c) operates some sharpening too, but with a differential effect that favors locally "horizontal" trajectories as a result of its built-in reassignment, which is achieved along the only frequency direction.

This becomes even more the case if we accept the "δ-model"

$$\rho(t, \omega) = \sum_{n=1}^{N} a_n^2(t)\, \delta(\omega - \omega_n(t)), \qquad (10.17)$$

in which $a_n(t)$ and $\omega_n(t)$ stand for the instantaneaous amplitudes and frequencies, respectively, of the N individual chirp components, as a model for multicomponent chirp signals. Following [99], this idea of sparsity offers a new perspective for constructing a time-frequency distribution that would approach the idealized model (10.17).

To capture the rationale of this approach, it is worth returning to Cohen's class (6.17) and understanding the role played by its weighting function $\varphi(\xi, \tau)$ from a geometric point of view. As already explained, the Wigner distribution is the 2D Fourier transform of the ambiguity function, which has the status of a time-frequency correlation. A correlation function is a measure of similarity (in the L_2 sense) between a signal and its shifted versions, thus guaranteeing that its magnitude is maximum at the origin (no shift), since a signal cannot resemble more to another signal than to itself! Were the signal be to made of two shifted versions of one given waveform, it is clear that its correlation would consist of three terms: one "central" contribution around the origin, corresponding to the superposition of the *autocorrelations* of the two components, and two additional "side" contributions, symmetrical and located at a distance from the origin that equals their shift, corresponding to the two *cross-correlations* between the two components. This is illustrated in Figure 10.8, which parallels Figure 9.1, complementing the Wigner distributions by the corresponding ambiguity functions.

> What we learn from this brief analysis is the origin of the terms of the Wigner distribution: they are the Fourier image of the *cross-ambiguity* terms that lie away from the origin of the plane.

This is exactly where Cohen's class enters the game for helping in the removal of such undesirable interference components. Indeed, it follows from (6.17) that the construction of distributions of Cohen's class is pretty much similar to that of the Wigner distribution, except that a weighting is applied to the ambiguity function prior to the (inverse) Fourier transformation. This offers a key for removing interference terms: it suffices to make use of a weighting function that vanishes away from the origin of the plane, in those domains where the cross-ambiguities are located. If this vanishing occurs in all directions in the ambiguity plane, the weighting function is "low-pass," inducing a smoothing on the Wigner distribution in the time-frequency plane. This is exactly what happens with the spectrogram, whose Cohen's weighting function is given by $\varphi(\xi, \tau) = A_h^*(\xi, \tau)$ in the general case of an arbitrary window $h(t)$, and therefore by the 2D Gaussian (6.9) when $h(t)$ is chosen as the circular Gaussian $g(t)$ defined in (9.1).

This perspective on interference terms reduction has prevailed for decades since the early introduction of its rationale [96], and it led to many variations and a flourishing activity in the 1990s [82]. However, as illustrated in Figure 10.8, this approach permits a reduction of interference terms that necessarily goes along with a smearing of the signal terms. This is because it basically relies on the Fourier transform, which has to face uncertainty. In the example of Figure 10.8, we see that, when the two shifted components are spaced sufficiently far apart, a spectrogram weighting/smoothing simply cancels out their interference, though at the expense of smearing the signal terms (see Figure 9.1(b) and (e)). When the two components get closer, however, the cross-terms tend to overlap with the auto-terms in the ambiguity plane, and their removal would call

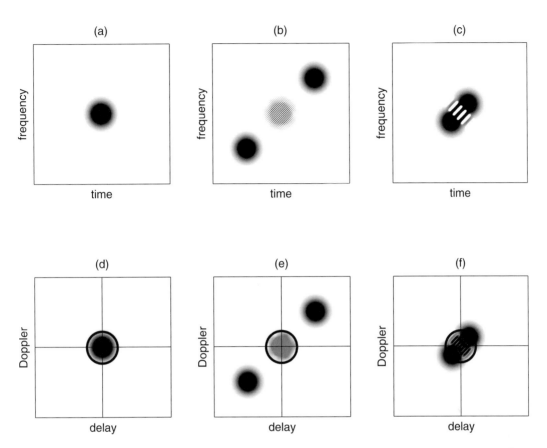

Figure 10.8 (This figure is is a complement to Figure 9.1). Wigner distributions (top row) and ambiguity functions (bottom row) for one logon (left column) and two logons (middle and right columns). Looking at (b) and (e) where the two components are far apart, we see that the interference term of the Wigner distribution is the Fourier image of the two cross-ambiguity terms that lie away from the origin. The same mechanism holds in (c) and (f) where the two components get closer, but the cross-ambiguity terms overlap in this case with the auto-ambiguity terms that surround the origin, making it impossible to disentangle them by a Gaussian spectrogram weighting whose essential support is given by the black circle. Note that the time and frequency axes in (a)–(c) are scaled by a factor 2 as compared to (d)–(f).

for a more concentrated weighting function in this ambiguity plane, which would in turn increase the smearing in the time-frequency plane.

The situation could seem desperate, but one way out is possible if we assume that the distribution we seek is *sparse* in the time-frequency plane. From this perspective, rather than making use of the Fourier transform (which amounts to giving an L_2 approximation to the Wigner distribution from the partial knowledge of the ambiguity function after weighting), we can think of enforcing sparsity in the time-frequency domain by adopting an L_1 setting, in the spirit of "compressed sensing" approaches [97, 98]. More precisely [99],

> Considering that the L_1 norm is an adequate measure of sparsity, the idea is to find the *sparsest* time-frequency distribution whose Fourier transform coincides with the ambiguity function within a domain that is centered around the origin and small enough to get rid of cross-terms.

In other words, if we use $\hat{\rho}_x(t,\omega)$ to denote the targeted time-frequency distribution and $\varphi_D(\xi,\tau)$ to denote a weighting function defined over a domain D around the origin of the ambiguity plane so as to get rid of most cross-terms, the "classical" approach

$$\hat{\rho}_x = \mathbf{F}^{-1}\{\varphi_D A_x\} \tag{10.18}$$

is replaced by the optimization-based solution:

$$\hat{\rho}_x = \arg\min_\rho \|\rho\|_1, \quad \text{s.t.} \quad \mathbf{F}\{\rho_x\} = A_x|_{(\xi,\tau)\in D}, \tag{10.19}$$

where $\mathbf{F}$ stands for the Fourier transform from (t,ω) to (ξ,τ), and $\mathbf{F}^{-1}$ for its inverse. Proceeding this way, it turns out that [99, 100]:

> The domain D can be dramatically smaller than the essential support of the ambiguity function of a logon, thus allowing for both an effective rubbing out of interferences and a sharp localization of the distribution on the ideal time-frequency structure.

This is illustrated in Figure 10.9, where we deliberately choose a toy signal made of the superimposition of a Hermite function (of order 25) and a sinusoidally frequency-modulated chirp, with a time-frequency signature thus consisting of about one period of a sine wave within a circle. This signal is unlikely to be encountered in nature, but it is also a typical example of a waveform whose structure would be hard to disentangle without considering time and frequency jointly. As can be observed in Figure 10.9(d), the sparse solution essentially captures the actual time-frequency structure of the analyzed signal, although its computation involves a subset of the ambiguity function that is extremely small (see Figure 10.9(b)). Using this domain to support a weighting function would have resulted in an overspread distribution with the conventional (inverse) Fourier method, whereas the assumption of sparsity – which clearly makes sense in this simple example – enables us to overcome the time-frequency trade-off "localization versus interference."

Remark. In contrast with usual "compressed sensing" approaches to inverse problems (see, e.g., [98]), it is important to understand that the goal here is not to recover the whole inverse Fourier transform of a function (in this case, the ambiguity function) from a small set of measurements, since this would result in the Wigner distribution itself! The objective is rather to get access to a hypothetical object whose time-frequency structure is essentially dictated by the local behavior of the ambiguity function in the vicinity of the origin of the plane. What is shown in [99] is that a "too-small" domain around this origin cannot capture the necessary information about auto-terms, while a "too-large"

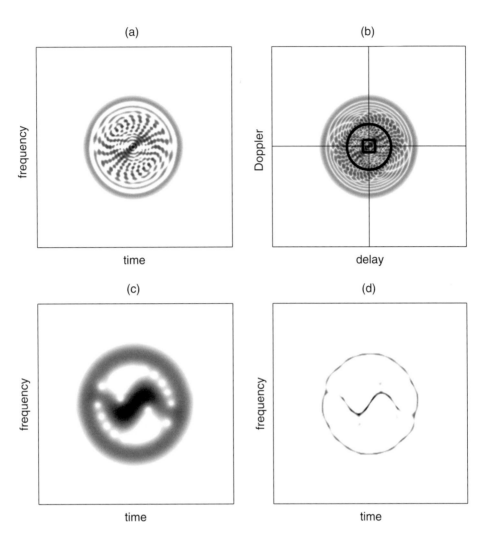

Figure 10.9 Sparse time-frequency distributions. The toy signal considered in this figure is made of the superimposition of a Hermite function (of order 25) and a sinusoidally frequency-modulated chirp. The time-frequency signature of this signal, which consists of about one period of a sine wave within a circle, is overwhelmed by interferences in the Wigner distribution (a) and blurred in the Gaussian spectrogram (c). The sub-diagram (b) plots the ambiguity function, with the essential support of the spectrogram weighting function indicated by the black circle. The much smaller black square corresponds to the domain D used for the optimization program (10.19), with (d) as the resulting sparse time-frequency distribution.

one incorporates a contamination with undesirable signatures of cross-terms. What is striking is that the transition from "small" to "large" is observed when the domain in the ambiguity plane is chosen to have an area of the order of a cell with minimum time-frequency uncertainty.

While we refer interested readers to [99] or [100] for details about the method, its interpretation, and how to actually solve the optimization problem (10.19), we will conclude this section with one additional pro and two cons for this sparsity-based approach to time-frequency distributions. The additional pro is that, even if this is not entered into the problem as an explicit constraint, the solution is generally almost nonnegative everywhere [100], making it a sharpened and almost interference-free spectrogram-like distribution. The first con is that the overall procedure is computationally expensive and hardly feasible for "not-so-small" signals. The second one is that, while the representation it offers is visually attractive, it is unfortunately not invertible. We will therefore not go into further detail here, although it did seem interesting to point out such an alternative to reassignment, whose objectives are quite similar.

10.5   Wedding Sharpening and Reconstruction

In the introductory part of this chapter, we invoked the "Holy Grail" of time-frequency analysis as the method that would lead to a distribution both sharply localized and free from interference, i.e., such that its outcome would approach at best the "δ-model" (10.17) in the case of multicomponent chirp signals. Phrased this way, however, this only concerns *visualization* and, in the light of what we have seen in the two previous sections, we can now be more demanding and ask in addition for a possible *reconstruction* of the localized components.

> Reassignment and synchrosqueezing only go halfway in this twofold, more ambitious direction – the former favoring localization at the expense of invertibility, and the latter sacrificing isotropic sharpening for reconstruction.

It therefore becomes legitimate to explore approaches that would ideally combine both requirements. Without going into too much detail, we will mention here that another possibility that has been recently explored [101] as a variation on synchrosqueezing or reassignment – one that aims to improve upon their standard formulations.

As for synchrosqueezing, it is easy to understand that a sharp localization is achieved for (local) *tones* since, in such a case, the actual direction of reassignment is precisely parallel to the frequency axis at each time, with the frequency of the tone being the fixed point of the operator. In this very specific case, synchrosqueezing is doubly successful (in terms of localization and reconstruction) since the only information that is necessary to consider is attached to the current time t. Whenever the (local) tone is replaced by a (local) chirp, however, this ceases to be true. Indeed, the frequency reassignment is only one component of a vector which, in this case, also contains a nonzero reassignment component in time. Neglecting this contribution by reassigning in frequency only results in a bias that limits localization, yet does not affect reconstruction.

As for reassignment now, we could imagine mimicking synchrosqueezing by working with the STFT $F_x(t, \omega)$ instead of the spectrogram $S_x(t, \omega)$, while applying the recipe (10.1) with the necessary changes. The problem in this case is that no direct "vertical" reconstruction is possible, since the values to add up are not synchronous when the two components of the reassignment vector are both nonzero. One solution has been proposed in [101] in the case of chirps whose instantaneous frequency $\omega_x(t)$ is (locally) linear. Given some evaluation point (t, ω), the rationale is to move a suitably *phase-shifted* version of the actual SFTF value $F_x(t, \omega)$ according to:

$$\tilde{F}_x(t, \omega) = \iint_{-\infty}^{\infty} F_x(\tau, \xi) \, \exp\{i\Psi_x(\tau, \xi; t, \omega)\} \, \delta\left(t - \hat{t}_x(\tau, \xi), \omega - \hat{\omega}_x(\tau, \xi)\right) \, d\tau \frac{d\xi}{2\pi},$$

(10.20)

in which the phase is made dependent on an estimate

$$\hat{\alpha}(t, \omega) = \frac{\partial \hat{\omega}_x(t, \omega)/\partial t}{\partial \hat{t}_x(t, \omega)/\partial t}$$

(10.21)

of the slope $\alpha(t, \omega)$ of the (locally) linear chirp.

This is possible because, assuming the (local) linearity of $\omega_x(t)$, the introduced phase term makes the straight line "slide" along itself, so that reassigning the phase-shifted STFT value $\hat{\alpha}(t, \omega)$ to $(\hat{t}_x(t, \omega), \hat{\omega}_x(t, \omega))$ results in reassigning the actual STFT value $F_x(t, \omega)$ to $(t, \omega_x(t))$.

Remembering that reassignment perfectly localizes linear chirps, we can then get an approximate reconstruction by simply evaluating this newly reassigned STFT along the instantaneous frequency line:

$$x(t) \approx \tilde{F}_x(t, \omega_x(t)).$$

(10.22)

Remark. Extensions of STFT synchrosqueezing have been proposed to both improve upon localization and guarantee reconstruction. This is, for instance, the case of the "high-order synchrosqueezing" introduced and discussed in [102].

11 A Digression on the Hilbert–Huang Transform

Reassignment and synchrosqueezing can be categorized as *postprocessing* approaches in the sense that a standard time-frequency transform (STFT and/or spectrogram) is first computed, while the sharp distribution, its disentanglement into distinct components, and the reconstruction of the corresponding waveforms, are all achieved in a second step on the basis of those computed values.

In contrast with such postprocessing approaches, a time-frequency description such as the one given by the "δ-model" (10.17) would be trivial in the case where some *preprocessing* could first disentangle the analyzed signal in its components: it would basically amount to *demodulating* each of those components in order to extract their instantaneous amplitudes and frequencies! Such an approach was actually proposed in the late 1990s, bypassing the need to resort to time-frequency analysis as a primary step and working directly in the time domain.

> This method, which is referred to as the *Hilbert–Huang Transform* [103], is composed of two distinct operations. The first operation is aimed at decomposing the signal into "modes," while the second one demodulates those modes and assigns them to the time-frequency plane according to (10.17).

A thorough discussion of the Hilbert–Huang Transform would require an extremely long digression in our time-frequency journey, so we will [briefly] summarize some of its key features here.

11.1 Empirical Mode Decomposition

Concerning the first operation, the data-driven decomposition is achieved by means of the so-called "Empirical Mode Distribution" (EMD) [104] whose rationale is to consider "any" complicated oscillatory waveform as being made of a fast oscillation on top of a slower one ... which can itself be decomposed the same way, and so on:

$$
\begin{aligned}
x(t) &= d_1(t) + r_1(t) \\
&= d_1(t) + d_2(t) + r_2(t) \\
&= \dots \\
&= \sum_{k=1}^{K} d_k(t) + r_K(t),
\end{aligned}
\tag{11.1}
$$

with $\{d_k(t), k = 1, \ldots K\}$ the extracted modes (the so-called "*Intrinsic Mode Functions*" (IMF) in Huang's terminology [103, 104]) and $r_K(t)$ the residual at some level K that can be fixed by the data and/or the user.

In a nutshell, EMD's motto can be summarized as:

> signal = fast oscillation + residual
>
> &
>
> iteration on residual

The key question is of course to define what is "fast" as compared to what is "slow." Looking at it from a purely formal viewpoint, the procedure described in (11.1) has the flavor of a *wavelet decomposition* [20], with the modes $d_k(t)$ emulating "detail" sequences, and residuals $r_k(t)$ the corresponding "approximations."

As is well known, wavelets operate this "detail versus approximation" dichotomy by *half-splitting* the frequency band, with details resulting from the high-pass (and approximations from the low-pass) filtering.

> The situation is different in the EMD case, where the wavelet idea of a predetermined filtering aimed at a separation based on the global frequency content is replaced by a data-driven approach making use of local oscillations.

More precisely, the EMD disentanglement "fast versus slow" is achieved by iterating a nonlinear operator, which acts at the level of an adaptive time scale that is defined by the distance between successive *local extrema*. When applied to one of the residuals $r_k(t)$, this *sifting* operator $\mathbf{S}$ is defined by the following procedure:

1. Identify the extrema of $r_k(t)$;
2. Interpolate with a cubic spline between the minima, so as to define an "envelope" $e_{\min}(t)$ (Do similarly between the maxima so as to define an "envelope" $e_{\max}(t)$);
3. Compute the mean envelope $m(t) = (e_{\min}(t) + e_{\max}(t))/2$;
4. Subtract this envelope to the initial waveform: $(\mathbf{S}r_k)(t) = r_k(t) - m(t)$.

When iterating M times this procedure, modes and residuals at scale $k + 1$ (or, equivalently, details and approximations at the same scale) are defined as $d_{k+1}(t) = (\mathbf{S}^M r_k)(t)$ and $r_{k+1}(t) = r_k(t) - d_{k+1}(t)$.

Remark. As it is clear from its name, the "Empirical Mode Decomposition" is essentially an *empirical* method that lacks a firm theoretical foundation. It is precisely when noticing this limitation that Ingrid Daubechies reintroduced the idea of synchrosqueezing as an alternative with much better controlled properties [105]. This is an interesting example of the revival of an ancient idea thanks to the arrival of a new one.

11.2   Huang's Algorithm

From a practical point of view, the overall EMD algorithm applied to a signal $x(t)$ can be given the following pseudo-code form:

$r_0(t) = x(t)$
for $k = 0 : K$ **do**
$\quad$ $r(t) = r_k(t)$
$\quad$ **for** $m = 1 : M$ **do**
$\quad\quad$ $d_{k+1}(t) = (Sr)(t)$
$\quad\quad$ $r(t) \leftarrow d_{k+1}(t)$
$\quad$ **end**
$\quad$ $r_{k+1}(t) = r_k(t) - d_{k+1}(t)$
end

One can wonder why it is necessary to introduce in the algorithm an inner loop that iterates the sifting process on a "proto-mode" $r(t)$ until it is accepted as an actual mode. The reason is to be found in the interpretation that is attached to what should be an acceptable mode (i.e., an "IMF"). In Huang's approach, an IMF should capture – as much as possible – the local oscillation, which means that it should be *locally zero-mean*. In general, this is not achieved when the operator S is applied only once, but – although no convincing proof has been established – this tends to become true when the signal is repeatedly sifted. Stopping sifting is more an art than a technique, with M being possibly fixed a priori by the user (a typical recommendation is to take $M \approx 10$), or determined by a desired degree of approximation of the zero-mean condition (see, e.g., [106] for details).

> The advantage of achieving zero-mean modes is that they are automatically amenable to the method's second operation, i.e., demodulation.

11.3   The Hilbert–Huang Transform

Indeed, in the case of multicomponent chirp signals such as those targeted by the "δ-model" (10.17), we can expect to recover the different components as IMFs of the form:

$$d_k(t) = a_k(t) \cos\{\varphi_k(t)\}. \tag{11.2}$$

Advocating the standard approach to demodulation based on the *Hilbert transform*, we can accept the quantities $a_k(t)$ and $\omega_x(t) = d\varphi_x(t)/dt$ as a reasonable approximation to the instantaneous amplitude and frequency, respectively. This permits the construction of a sort of time-frequency energy distribution, which reads:

$$H(t, \omega) = \sum_{k=1}^{K} a_k^2(t)\, \delta(\omega - \omega_k(t)), \tag{11.3}$$

with the overall method referred to as the *Hilbert–Huang Transform* [103].

In general, there is no reason that $K = N$, nor that (some of) the IMFs identify to the actual components. Typically, a signal of sample size T whose spectrum is strictly positive is decomposed by EMD in about $\log_2 T$ IMFs, and a necessary (but not sufficient) condition for directly recovering the signal components as some of the IMFs is that the former coexist for all times in the "δ-model" (10.17), and that their time-frequency trajectories $\omega_k(t)$ are well separated in frequency.

11.4   Pros, Cons, and Variations

To conclude this quick tour, we can consider a simple two-component example which consists of two chirping components, overlapping in both time and frequency, and embedded in white Gaussian noise, with a very high signal-to-noise ratio.

The decomposition itself is given in Figure 11.1, while the resulting "time-frequency distribution" is displayed in Figure 11.2, together with the corresponding Gaussian

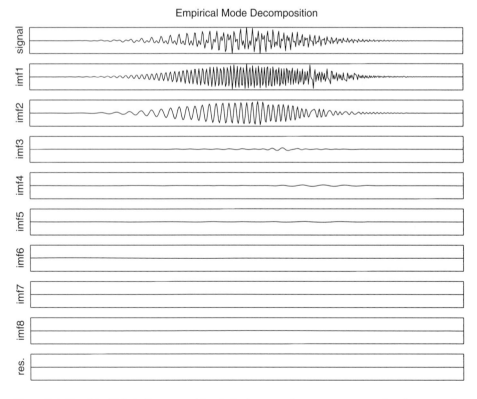

Figure 11.1 Empirical Mode Decomposition 1. In the case of a two-component signal composed of two chirping components embedded in white Gaussian noise, with a very high signal-to-noise ratio, EMD results in a collection of IMFs and one residual. In the present situation, the first two IMFs recover the two actual components, while the extra modes, due to noise and sampling, are negligible.

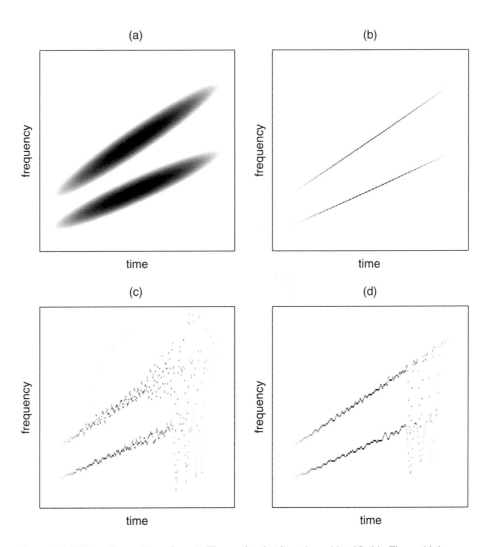

Figure 11.2 Hilbert–Huang Transform 1. The modes that have been identified in Figure 11.1 can be demodulated via the Hilbert method, thus permitting the creation of a time-frequency distribution by localizing the local modes energy along the instantaneous frequencies. This corresponds to the sub-diagrams (c) and (d), which differ in terms of estimation of the instantaneous frequency (crude difference operator in the first case and maximum likelihood in the second one). The sub-diagrams (a) and (b) display respectively the corresponding Gaussian spectrogram and its reassigned version, for sake of comparison.

spectrogram and its reassigned version, for the sake of comparison. In fact, two versions of the Hilbert–Huang Transform are presented, depending on the way the instantaneous frequency is estimated, either with a simple difference operator aimed at approaching the derivative (Figure 11.2(c)), or with a refined maximum likelihood estimator (Figure 11.2(d)).

If we are interested in obtaining the individual components, we could think of just making use of EMD and forgetting about the Hilbert part, which is much more unstable in terms of visualization. Unfortunately, since the extracted IMFs do not necessarily coincide with the actual components, a global time-frequency display is not to be neglected since it provides the user with better-structured information.

This is shown in Figures 11.3 and 11.4, in a case which illustrates the so-called *mode mixing effect*, i.e., the situation where one component is present in more than one IMF. The example considered in these figures is the same as before, except that the high-frequency chirp is stopped earlier than the low-frequency one. This results, by construction, in the splitting of the tone between the first IMF (after the high-frequency chirp is switched off) and the second IMF (when the local fastest oscillation is that of the chirp). Directly considering either mode individually would make no sense, while the time-frequency distribution recovers the continuity of the tone component (though with some fake extra contribution).

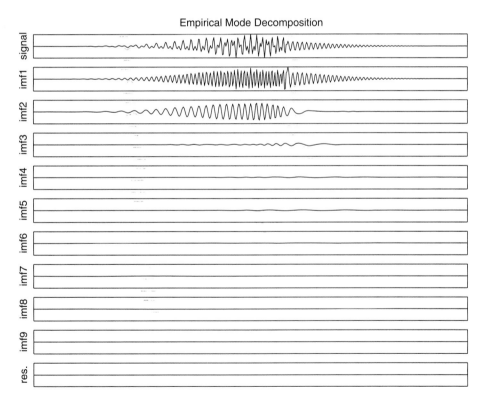

Figure 11.3 Empirical Mode Decomposition 2. In the case of an example similar to that of Figure 11.1, except that the high-frequency chirp is stopped earlier than the low-frequency one, EMD again results in a collection of IMFs and one residual. In this new situation, however, the low-frequency chirp is split in the first two IMFs (mode mixing effect), thus calling for some further processing in order to correctly identify the actual components.

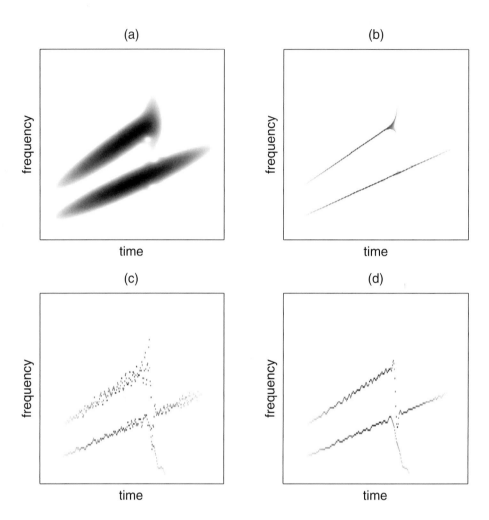

Figure 11.4 Hilbert–Huang Transform 2. The modes which have been identified in Figure 11.3 can be demodulated via the Hilbert method, thus permitting to create a time-frequency distribution by localizing the local modes energy along the instantaneous frequencies. This corresponds to diagrams (c) and (d), which differ in terms of estimation of the instantaneous frequency (crude difference operator in the first case and maximum likelihood in the second one). Diagrams (a) and (b) display the corresponding Gaussian spectrogram and its reassigned version, respectively, for the sake of comparison.This display of the signal energy permits us, to some extent, to recover the continuity of the low-frequency chirp that has been split over two IMFs, though with some fake extra contribution.

Remark. "Mode mixing problem" also occurs in noisy data, when the noise fluctuations have a time-varying variance. This has led to a flourishing development of variations in EMD. Most of them are *noise-assisted*, with the adding of some controlled level of noise to the data [107]. The rationale is that different realizations of added noise will make the mode mixing appear at different times. Computing such different decompositions and averaging them is expected to reinforce the "coherent" signal part and to cancel out the "incoherent" noise fluctuations.

As we did in Chapter 5 with the Wigner distribution, let us close this chapter with some "golden triangle" considerations regarding Empirical Mode Decompositions. EMD was essentially pushed forward by NASA oceanographer Norden E. Huang and his collaborators in the late 1990s as a way of disentangling a waveform made of many oscillations into "intrinsic" components [104]. The rationale is clearly based on physical intuition, reasoning in terms of local oscillations rather than Fourier modes or wavelets, so as to encompass possible nonstationarities and/or nonlinearities. Furthermore, thanks to the data-driven way oscillations are identified and extracted, the method is naturally adaptive. All of this makes EMD very appealing, and this is reinforced by the fact that it is quite easy to implement, this time establishing a link between physics and informatics in the "golden triangle."

The *caveat*, however, is that EMD strongly resists any mathematical analysis, at least in its original formulation. The method is not defined as a transform (despite the name "Hilbert–Huang Transform" that has, unfortunately, been coined for the whole process described in the previous chapter), but is defined only by an algorithm, analysis of which proves difficult. Two interlinked loops are involved, with stopping criteria (as well as other tuning parameters) whose more or less *ad hoc* choices are left to the user: as it is clear from its name, EMD is basically *empirical*.

The problem with the lack of a solid mathematical foundation is that two main options can be envisioned if one wants to go beyond a mere (and blind) application of the existing algorithm so as to get some confidence in the obtained results. One can either follow the EMD recipe and try to learn about the kind of information that can (or cannot) be retrieved in the case of extensive, well-controlled simulations, or one can keep the "spirit" of the method while modifying its realization and implementation so as to get some mathematical control (this is typically what can be done with synchrosqueezing and the like). Regardless of the approach, we see and must accept that the missing links between the mathematics summit and the other two summits in the "golden triangle" are clearly an obstacle to a wide acceptance of EMD by the signal processing community, in which well-established methodologies matter as much as applications.

12 Spectrogram Geometry 2

12.1 Spectrogram, STFT, and Bargmann

After having spent some time, for the sake of interpretation, with spectrograms considered as smoothed Wigner distributions, let us return to their more conventional definition as squared STFTs, i.e., (6.2) with (6.1). In both expressions, we maintain the explicit mention of the window $h(t)$ which, while still assumed to be real-valued for simplicity, is not necessarily circular Gaussian at this point.

With this definition at hand, we can now make a twist concerning the interpretation of the time-frequency plane.

> Classically, time and frequency are seen as Cartesian coordinates of the real plane, but they can equivalently be viewed as components of a complex-valued vector, thus identifying the time-frequency plane with the complex plane.

To be more precise, we can introduce the complex variable $z = \omega + it$, with the result that $t = (z - z^*)/2i$ and $\omega = (z + z^*)/2$ or, in other words, that the STFT can be expressed as a function of z and z^*. In the general case of an arbitrary window $h(t)$, we can therefore choose to rewrite (6.1) using the factorized form

$$F_x^{(h)}(t, \omega) = \mathcal{F}_x^{(h)}(z, z^*) \, \exp\left\{-|z|^2/4\right\}. \tag{12.1}$$

This could be thought of as an arbitrary factorization, but its rationale is to be found in what happens when we choose the circular Gaussian window $g(t)$ (9.1) – which has been of such a great importance in the previous discussions – for $h(t)$. In this case, it turns out that $\mathcal{F}_x^{(h)}(z, z^*)$ only depends on z [108]. Simplifying the notation, we can thus write

$$F_x(t, \omega) = \mathcal{F}_x(z) \, \exp\left\{-|z|^2/4\right\}, \tag{12.2}$$

assuming, as mentioned in the introduction of Chapter 9, that no superscript means "$(h) = (g)$."

In this way, $\mathcal{F}_x(z)$ becomes an *analytic* function that reads

$$\mathcal{F}_x(z) = \int_{-\infty}^{+\infty} \pi^{-\frac{1}{4}} \exp\{-s^2/2 - isz + z^2/4\} \, x(s) \, ds \tag{12.3}$$

and thus exactly identifies with the so-called *Bargmann transform* of the signal $x(t)$ [109].

The analyticity of $\mathcal{F}_x(z)$ has a number of consequences. Some of them will be detailed and used further in the following chapters. Meanwhile, we can draw first conclusions about the phase-magnitude relationships analyticity induces on the STFT [110]. Since, by construction, a STFT is a complex-valued representation, its complete characterization requires the knowledge of either its real and imaginary parts, or its magnitude and phase. Focusing on magnitude and phase, if we write

$$F_x(t, \omega) = M_x(t, \omega) \exp\{i\Phi_x(t, \omega)\}, \qquad (12.4)$$

we have, of course, $M_x(t, \omega) = |F_x(t, \omega)|$. If we rewrite (12.2) as

$$\mathcal{F}_x(z) = F_x(t, \omega) \exp\{(t^2 + \omega^2)/4\}, \qquad (12.5)$$

we can also define

$$\mathcal{M}_x(t, \omega) = |\mathcal{F}_x(z)| = M_x(t, \omega) \exp\{(t^2 + \omega^2)/4\}, \qquad (12.6)$$

while the phases of both $F_x(t, \omega)$ and $\mathcal{F}_x(z)$ are identical. Taking the logarithm of $\mathcal{F}_x(z)$ so as to deal directly with phase and (log-)magnitude, analyticity can then be transformed into Cauchy conditions that read [110]:

$$\frac{\partial \Phi_x}{\partial t}(t, \omega) = \frac{\partial}{\partial \omega} \log \mathcal{M}_x(t, \omega); \qquad (12.7)$$

$$\frac{\partial \Phi_x}{\partial \omega}(t, \omega) = -\frac{\partial}{\partial t} \log \mathcal{M}_x(t, \omega). \qquad (12.8)$$

We see, therefore, that choosing a circular Gaussian as a window for the STFT results in a strong coupling between phase and magnitude. This expresses the redundancy that is inherent to any STFT, but in a particularly simple – yet different – form.

This was previously formulated in terms of the reproducing kernel of the analysis, which was itself rooted in the Fourier uncertainty. This is now expressed in terms of magnitude and phase, but the overall conclusion is of the same nature: no independence between time and frequency, no pointwise interpretation of a time-frequency transform!

12.2 Reassignment Variations

We have seen previously that reassignment, as defined by (10.1), amounts to *moving* spectrogram values from the time-frequency point (t, ω) where they have been computed to another time-frequency point $(\hat{t}(t, \omega), \hat{\omega}(t, \omega))$ that is more representative of the local energy distribution. This action therefore defines a *vector field*

$$\hat{\mathbf{r}}_x(t, \omega) = \begin{pmatrix} \hat{t}(t, \omega) - t \\ \hat{\omega}(t, \omega) - \omega \end{pmatrix} \qquad (12.9)$$

that summarizes all moves taking place in the plane. If we confine ourselves to the circular Gaussian case, and if we make use of the phase-based expressions (10.8)–(10.9) of the reassignment operators, it follows from the phase-magnitude relationships (12.7)–(12.8) that

$$\hat{t}(t, \omega) - t = -\frac{t}{2} + \frac{\partial}{\partial t} \log \left(|F_x(t, \omega)| \, \exp\{(t^2 + \omega^2)/4)\} \right)$$

$$= \frac{\partial}{\partial t} \log |F_x(t, \omega)| \tag{12.10}$$

whereas, proceeding similarly with the frequency reassignment, we have:

$$\hat{\omega}(t, \omega) - \omega = \frac{\partial}{\partial \omega} \log |F_x(t, \omega)|. \tag{12.11}$$

Combining these two identities, we end up with the particularly simple result [108]:

$$\hat{\mathbf{r}}_x(t, \omega) = \nabla \log |F_x(t, \omega)| = \frac{1}{2} \nabla \log S_x(t, \omega), \tag{12.12}$$

which expresses that:

> The reassignment vector field is the gradient of a scalar potential function that happens to be the log-magnitude of the STFT (or of the log-spectrogram).

Remark. In its most general form, reassignment of a spectrogram with a window $h(t)$ requires, besides the computation of the STFT with that window, the computation of two additional STFTs with windows derived from $h(t)$, namely $(\mathcal{T}h)(t) = t.h(t)$ and $(\mathcal{D}h)(t) = (dh/dt)(t)$. It has already been remarked that, in the case of a Gaussian window, only one of these two STFTs is necessary, since the two windows are proportional. Furthermore, if the Gaussian window is circular, there is even no need – at least in principle – to compute any additional STFT, since the entire information about the reassignment vector field can be derived from simply having knowledge of the ordinary spectrogram.

Variation 1 – Differential reassignment. As first mentioned in [108], this suggests revisiting the reassignment process in terms of a dynamic system whose behavior would be governed by the log-spectrogram as a potential function. From this perspective, the spectrogram value at a given time-frequency can be viewed as a particle, while the reassignment vector field would be the velocity field controlling its motion. In other words, one can think of a *differential reassignment* whose realization amounts to integrating the motion equations

$$\frac{dt}{d\tau}(\tau) = \hat{t}(t(\tau), \omega(\tau)) - t(\tau); \tag{12.13}$$

$$\frac{d\omega}{d\tau}(\tau) = \hat{\omega}(t(\tau), \omega(\tau)) - \omega(\tau), \tag{12.14}$$

with the initial conditions $(t(0), \omega(0)) = (t, \omega)$, and that integration continues until convergence to a maximum of the log-spectrogram.

Variation 2 – Adjustable reassignment. In contrast to this continuous evolution, we can come back to (10.8) and (10.9) which, when combined with (12.9), leads to

$$\hat{\mathbf{r}}_x(t, \omega) = -\mathbf{R}_x(t, \omega), \tag{12.15}$$

with

$$\mathbf{R}_x(t, \omega) = \left(\begin{array}{c} t/2 + \partial\Phi_x(t, \omega)/\partial\omega \\ \omega/2 - \partial\Phi_x(t, \omega)/\partial t \end{array} \right). \tag{12.16}$$

Thanks to (12.12), the *ridges* (i.e., those lines onto which the log-spectrogram is locally maximal) are fixed points of the reassignment operators, implying in turn that reassignment itself can be seen as the first iteration of a fixed point algorithm of the form

$$y_{n+1} = y_n - \lambda f(y_n); n \in \mathbb{N}, \lambda > 0, \tag{12.17}$$

which is aimed at converging to a root of $f(.)$. Among the various possibilities offered in choosing λ, [111] has proposed retaining the so-called *Levenberg-Marquardt* scheme, which is defined by

$$y_{n+1} = y_n - \frac{1}{f'(y_n) + \mu} f(y_n), \tag{12.18}$$

where $\mu > 0$ is a control parameter that permits some robustness (as compared to Newton's algorithm corresponding to $\mu = 0$), in addition to adaptivity (as compared to a fixed value). By following this idea, we can thus modify the basic reassignment equation (12.15) according to

$$\tilde{\mathbf{r}}_x(t, \omega) = -(\nabla\mathbf{R}_x(t, \omega) + \mu\mathbf{I})^{-1} \mathbf{R}_x(t, \omega), \tag{12.19}$$

where $\mathbf{I}$ stands for the identity matrix. Given (12.16), the right-hand side of this equation involves first-order, second-order, and mixed derivatives of the phase. Fortunately, as a generalization of the equivalence that holds between (10.8)–(10.9) on the one hand and (10.4)–(10.5) on the other hand, all of those derivatives can be more efficiently computed thanks to STFTs with suitably chosen windows [111]. Once the new displacements have been computed this way, the newly modified spectrogram follows from (10.1) as before, with $(\hat{t}(t, \omega), \hat{\omega}(t, \omega))$ replaced by $(\tilde{t}(t, \omega), \tilde{\omega}(t, \omega))$ derived from (12.19).

The main point of interest of this *Levenberg-Marquardt reassignment* is that, thanks to the tuning of the μ parameter, it allows for an adjustment of the degree of energy concentration in the modified spectrogram, beyond the binary choice "full reassignment" $(\mu \to 0)$ versus "no reassignment" $(\mu \to +\infty)$.

Figure 12.1 uses a simple example to illustrate the transition between the ordinary spectrogram and its usual reassigned version that is permitted by the Levenberg-Marquardt reassignment.

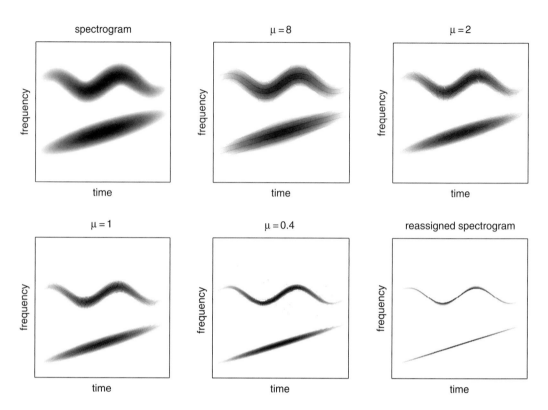

Figure 12.1 Levenberg-Marquardt reassignment. By varying the μ-parameter, the Levenberg-Marquardt reassignment permits us to control the time-frequency energy concentration, offering a continuous transition in between the ordinary spectrogram (top left) and its usual reassigned version (bottom right).

Remark. In the specific case of signals made of Gaussian waveforms, a suitably "scaled" modification of reassignment [112] can lead to a *perfect* localization that improves upon the shrinkage established in (10.6). This can be achieved by modifying (10.4)–(10.5) according to

$$\hat{t}(t, \omega) = t + c_t \operatorname{Re} \left\{ \frac{F_x^{(\mathcal{T}h)}(t, \omega)}{F_x(t, \omega)} \right\} \tag{12.20}$$

and

$$\hat{\omega}(t, \omega) = \omega - c_\omega \operatorname{Im} \left\{ \frac{F_x^{(\mathcal{D}h)}(t, \omega)}{F_x(t, \omega)} \right\}, \tag{12.21}$$

with c_t and c_ω, two control parameters that can be either matched to the analyzed signal when some a priori information is available, or otherwise estimated in a data-driven way so as to, e.g., minimize the Rényi entropy of the scaled reassigned spectrogram.

12.3 Attractors, Basins, Repellers, and Contours

Expression (12.12) tells us that, in theory, only the STFT magnitude carries all the phase information that is needed for reassignment. It also tells more about the way reassignment operates, namely that all vectors plot the direction of maxima of the magnitude.

> Those maxima appear therefore as *attractors* for the vector field, as well as *fixed points* since, thanks to (12.10) and (12.11), no move is undergone at those points.

An equivalent picture can be given by switching from the log-spectrogram $\log S_x(t, \omega)$ to its opposite $-\log S_x(t, \omega)$, and trading maxima for minima accordingly. This is illustrated in Figure 12.2 for the sum of the two chirp signals as in Figure 12.1 (one with a linear frequency modulation, the other with a sinusoidal one), both modulated in amplitude by a Gaussian. Plotted this way, the (negative of the) log-spectrogram resembles a double bowl whose gradient controls the direction of the reassignment vector field locally. Spectrogram values are therefore attracted by local minima (considering those values as particles, they tend to "fall' towards the bottom of the bowl to which they belong), and the reassigned distribution sums up all contributions originating from a time-frequency domain that can be naturally thought of as a *basin of attraction*.

When only one component is present, there is only one such basin, but as soon as we have more than one component, each of them has its own basin of attraction. Their co-existence defines a separatrix in between each pair of components such that, depending on whether a given time-frequency point is on one side or on the other, it will be attracted to a specific basin.

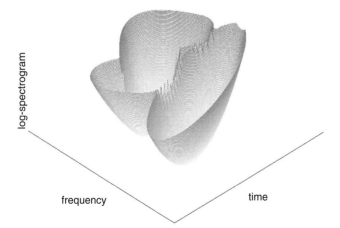

Figure 12.2 Potential function. The log-spectrogram (whose opposite value is plotted here) is the potential function whose gradient is the reassignment vector field. In the case of multicomponent signals, this corresponds to as many "bowls" as the number of components. Each of them can be viewed as defining a basin of attraction attached to the local extrema within the bowls.

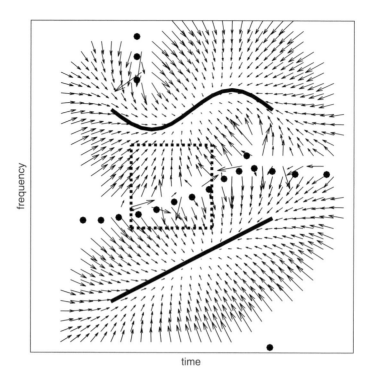

Figure 12.3 Attractors and separatrix. This figure displays the reassignment vector field corresponding to the potential function of Figure 12.2. It shows that reassignment vectors plot in the direction of local maxima whose locus is essentially supported by the time-frequency trajectories of the instantaneous frequencies, which appear as solid lines. Zeros of the spectrogram, which are indicated as dots, tend to align along the separatrix between the two basins of attraction defined by the vectors pointing towards a given set of maxima. A close-up view of the inner square appears in Figure 12.4.

The reassignment vector field corresponding to the potential function of Figure 12.2 is plotted in Figure 12.3. The arrows indicate the way all spectrogram values within a basin of attraction are reassigned to a much narrower domain, with a sharpened distribution as a by-product. In the case of the linear chirp, the reassignment is almost perfect (it would be if the duration were infinite, with no amplitude modulation), with all vectors ending on the line of the instantaneous frequency. Concerning the sinusoidal frequency modulation, the situation is pretty similar except when the curvature of the instantaneous frequency is too large to guarantee a quasi-linear approximation within the time-frequency window.

A question of great interest is, of course, the delineation of such basins of attraction. We will come back to this more precisely later in Chapters 14 and 15, but we can indicate now that two broad families of approaches are possible for handling this problem with the help of the reassignment vector field.

Concerning one first possibility, let us come back to Figure 12.3 which also displays the zeros of the spectrogram as full dots. What happens is that those zeros lie precisely along the separatrix between the two basins of attraction, while reassignment vectors originating from them plot opposite directions. This can be viewed as a consequence of the relation (12.12) which exists between the reassignment vector field and its associated potential function, when switching from local maxima to local minima. As a consequence, we can claim that:

> Zeros of the spectrogram are *repellers* for the reassignment vector field.

This is shown more clearly in Figure 12.4, which zooms in on Figure 12.3. Besides the zeros that are marked as full dots, *saddle points* are indicated as circles. By construction, such points appear between zeros that are repellers. They essentially lie along the separatrix too, as unstable extrema. Making use of zeros for basins of identification and signal disentanglement will be the topic of Chapter 15.

Now, as for the second approach, one more feature can be extracted from our knowledge of the reassignment vector field. It consists in going back to the general expressions (10.4)–(10.5) which, when the analyzing window $g(t)$ is circular Gaussian, can be expressed only by means of the ratio

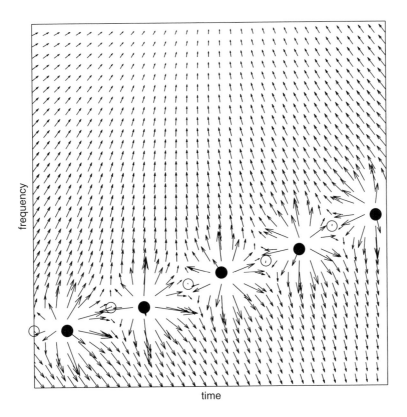

Figure 12.4 Repellers. Zooming in on Figure 12.3 shows that zeros of the spectrogram (indicated by the full dots) act as repellers for the reassignment vector field. Saddle points (marked as circles) appear between zeros and are essentially aligned along the separatrix between the basins of attraction.

$$R_x(t, \omega) = \frac{F_x^{(\mathcal{T}g)}(t, \omega)}{F_x(t, \omega)}, \tag{12.22}$$

since $(\mathcal{D}g)(t)$ happens to be proportional to $(\mathcal{T}g)(t)$ in this case. Fixed points of reassignment, i.e., those (t, ω) for which $\hat{t}_x(t, \omega) = t$ and $\hat{\omega}_x(t, \omega) = \omega$, are characterized by the vanishing of both $\mathrm{Re}\{R_x(t, \omega)\}$ and $\mathrm{Im}\{R_x(t, \omega)\}$. When taken independently, these two conditions are just special cases of the more general one:

$$\mathrm{Im}\{R_x(t, \omega)\exp\{-i\theta\}\} = 0, \tag{12.23}$$

with $\theta = 0$ and $\pi/2$, respectively. The expression (12.23) can be rephrased as the orthogonality condition:

$$\langle \hat{\mathbf{r}}_x(t, \omega), u_{\theta+\pi/2} \rangle = 0, \tag{12.24}$$

where u_λ stands for the unit vector of direction λ.

> The time-frequency points, such that (12.23) (or, equivalently, (12.24)) is satisfied, define time-frequency loci referred to as *contours* [113].

Contours have a number of specific properties [113, 115]. They first link together time-frequency points where the projected reassignment vector *changes sign*. This might be the case at local maxima along ridges, but this can also be encountered in valleys, as well as at saddle points which are either minima or maxima, depending on the chosen direction. Second, due to analyticity, contours never branch nor cross; they terminate at zeros and can form loops.

From 12.23, contours are curves defined as the locus of fixed points of a modified reassignment process that is forced to operate along the direction $\theta + \pi/2$ for a given orientation θ. They depend therefore on the chosen direction of projection for the reassignment vector, suggesting that the choice of this direction should be driven by the nature of the analyzed signal. In order to support this supposition, let us recall that, in the case of a circular Gaussian window, the analysis is isotropic; when the local value of a spectrogram at a given point (t, ω) is in the vicinity of a linear instantaneous frequency $\omega_x(s) = \omega_0 + \alpha s$, it is reassigned to a point $(\hat{t}(t, \omega), \hat{\omega}(t, \omega) = \omega_0 + \alpha \hat{t}(t, \omega))$ which is the *orthogonal projection* of (t, ω) on the instantaneous frequency $\omega_x(s)$, regardless of the slope α. In terms of contours, this means that the direction should be chosen as that of the linear chirp or, equivalently, as being orthogonal to the reassignment vector.

In the more general case of a chirp with a nonlinear instantaneous frequency, the local slope varies with time, but the argument about direction can be advocated locally [116, 117]. From a conceptual point of view, the reassignment vector can be deduced directly from the log-spectrogram surface (as its gradient), and contours can be derived from the change of sign when computing a directional gradient that is locally data-driven.

An illustration of what contours look like is given in Figure 12.5. The chosen example is similar to the one used throughout this chapter, with a small amount of noise superimposed (signal-to-noise ratio of 50 dB) so as to avoid numerical instabilities

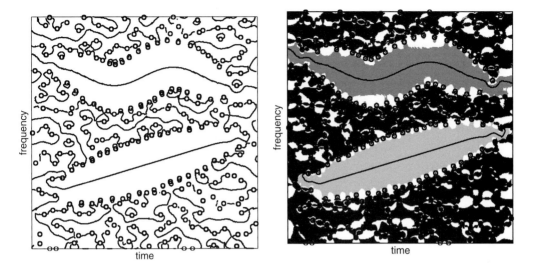

Figure 12.5 Contours and basins. Left: identifying the change in sign of reassignment vectors, in a given direction, permits us to single out points that define curves referred to as contours. In this example (which replicates the situation of the previous figures, with a small amount of noise superimposed), the direction is fixed locally on the basis of the reassignment vector. Contours define ridges along the instantaneous frequencies of the two chirp components, as well as curves that never branch nor cross, while passing through the zeros of the transform (indicated by circles). Right: labeling each time-frequency point by the contour that is closest to its reassignment target results in areas (here identified with different shades of gray) that corrrespond to basins attached to the different components.

due to very small values. What is revealed by this figure is the existence of long, smooth trajectories that essentially "draw" the instantaneous frequencies of the two chirp components. Those trajectories appear to be somehow in the middle of areas that are clear of any zero. Contours can be ordered by length and labeled, and each time-frequency point can then itself be labeled by the contour that is closest to its reassignment target. This is also a way of defining the *basins*, whose concatenation paves the entire time-frequency plane.

Remark. Contours have been a means to briefly mention the existence of *ridges*, which are essentially time-frequency trajectories that reflect the time-varying structure of chirps. Since the pioneering work [118], the idea of exploiting both amplitude and phase information for extracting such characteristic curves has been the subject of an abundance of literature. Having chosen here to focus on zeros instead, we will not comment further on such possibilities; we refer interested readers to, e.g., [119].

The examples considered so far call for a better knowledge of two broad categories of questions, namely the influence of noise on a spectrogram and the specific role played by its zeros. These are precisely the topics of the next chapters.

13 The Noise Case

Considering spectrograms in noise-only situations may serve different purposes. Indeed, time-frequency analysis is in large part a matter of geometry, in the sense of spatial organization of structured patterns in the plane. In this respect, since noise is supposed to be "unstructured," its study is expected to reveal features that are in some sense characteristic of disordered situations with no specific spatial organization. This makes statistics enter the picture, with the *a contrario* idea that the more we know about noise-only situations, the better we can say something about possibly embedded signals on the basis of a rejection of the (noise-only) null hypothesis.

So as to formalize "maximum disorder," we will consider here *white noise*, with the further assumption of Gaussianity. As was discussed in Chapter 3, speaking of "white Gaussian noise" still offers some degrees of freedom, depending in particular on whether it is real-valued or complex-valued. In most of the following cases, we will confine ourselves to the case of *complex white Gaussian noise* $n(t) \in \mathbb{C}$ which is characterized, as in (3.29)–(3.31), by the first- and second-order properties:

$$\mathbb{E}\{n(t)\} = 0; \tag{13.1}$$

$$\mathbb{E}\{n(t)\, n(s)\} = 0; \tag{13.2}$$

$$\mathbb{E}\{n(t)\, n^*(s)\} = \gamma_0\, \delta(t - s). \tag{13.3}$$

This will serve as an idealized model for more realistic situations (e.g., real-valued data or their analytic versions); interested readers are referred to [120] for a thorough investigation of the relevance of such a model.

13.1   Time-Frequency Patches

Before proceeding to a more precise analysis, let us get some insight about what time-frequency representations may look like in noise-only situations.

Figure 13.1 displays, for the sake of comparison, circular Gaussian spectrograms computed on four different realizations of complex white Gaussian noise. All diagrams are of course different, but they share the common feature of distributing, in some

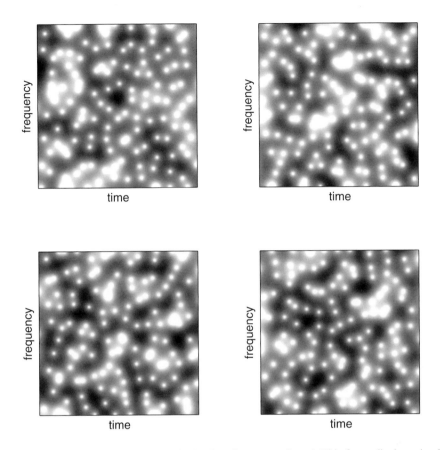

Figure 13.1 White Gaussian noise in the time-frequency plane 1. This figure displays circular Gaussian spectrograms computed on four different realizations of white Gaussian noise.

homogeneous way, *patches* where energy tends to locate, intertwined with a constellation of low energy values (indeed, zeros, as we will see).

According to (12.12), any of those (log-)spectrogram surfaces, made of bumps and troughs, can be viewed as the potential function whose gradient is the reassignment vector field. Within this picture, we have seen that while local maxima act as attractors, zeros are repellers. This results in a reassigned spectrogram made of highly concentrated regions in the vicinity of spectrogram local maxima, with voids in between, surrounding the zeros of the transform. This froth-like pattern – which had been noticed since the early developments of reassignment [114], and has been more recently highlighted in [115] – is illustrated in Figure 13.2.

It follows from those simple observations that spectrograms and their reassigned versions exhibit some specific structure that accounts for a time-frequency description of the inherent disorder of complex white Gaussian noise. We will now look more closely at this structure so as to later take advantage of the information it can provide for possibly identifying and extracting signals embedded in such a noise realization.

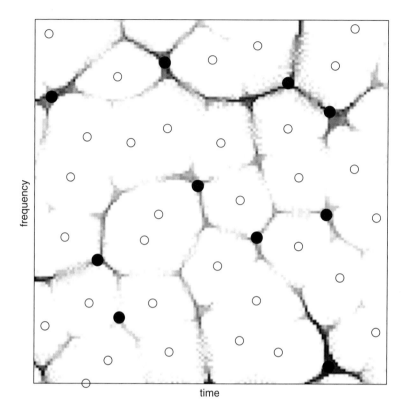

Figure 13.2 White Gaussian noise in the time-frequency plane 2. This figure displays the circular Gaussian reassigned spectrogram of some realization of white Gaussian noise, together with local maxima (full dots) and zeros (circles). It shows that reassigned values tend to concentrate in the vicinity of spectrogram local maxima, which run between the void regions surrounding the zeros of the transform.

13.2   Correlation Structure

As a function of the time difference $t - s$ only, (13.3) expresses that $n(t)$ is a stationary process, with γ_0 the level of its power spectrum density since, thanks to the Wiener-Khintchine-Bochner relation, we have:

$$\int_{-\infty}^{\infty} \mathbb{E}\{n(t)n^*(t - \tau)\} \exp\{-i\omega\tau\}d\tau = \gamma_0. \tag{13.4}$$

Of course, stationarity implies that each realization is of infinite duration (with an equal distribution at each time), with the consequence that none can be of finite energy, and that the direct computation of their Fourier transform is prohibited. Fortunately, a spectrogram involves a windowing prior to the Fourier transform, which allows us to formally manipulate the idealization of complex white Gaussian noise if we want to evaluate the statistical properties of its values.

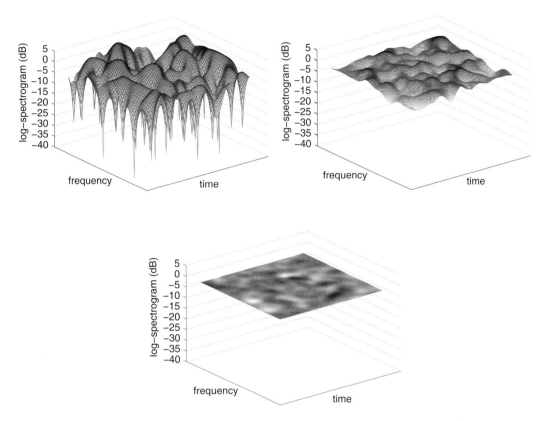

Figure 13.3 Average spectrogram. The top left surface displays the log-spectrogram of one realization of complex white Gaussian noise and its high variability. The top right surface displays the ensemble average of such log-spectrograms over 10 (and the bottom surface displays the ensemble average of such log-spectrograms over 5,000) independent realizations, illustrating the eventual convergence to a flat surface of level γ_0 (= 1 in the present case).

Making use of (13.3) and assuming, as usual, that the circular Gaussian window is of unit energy, we can first establish that:

$$\mathbb{E}\{S_n(t, \omega)\} = \gamma_0. \tag{13.5}$$

> The average spectrogram is flat and equal to the power spectrum density γ_0 for all times and all frequencies.

This expected result, which reflects the second order stationarity of complex white Gaussian noise, is illustrated in Figure 13.3.

Proceeding in the same way, we can evaluate the second-order statistical link that exists between any two spectrogram values at locations (t, ω) and (t, ω') in the time-frequency plane. Developing the covariance

$$\text{cov}\{S_n(t, \omega), S_n(t', \omega')\} = \mathbb{E}\{(S_n(t, \omega) - \gamma_0)(S_n(t', \omega') - \gamma_0)\} \tag{13.6}$$

as a quadruple integral, we end up with an expression involving fourth-order terms of the form $\mathbb{E}\{n(t_1)n^*(t_2)n(t_3)n^*(t_4)\}$ which, thanks to Isserlis' formula, can be developed into pair products as

$$\mathbb{E}\{n(t_1)n^*(t_2)n(t_3)n^*(t_4)\} = \mathbb{E}\{n(t_1)n^*(t_2)\}\,\mathbb{E}\{n(t_3)n^*(t_4)\}$$
$$+ \mathbb{E}\{n(t_1)n(t_3)\}\,\mathbb{E}\{n^*(t_2)n^*(t_4)\}$$
$$+ \mathbb{E}\{n(t_1)n^*(t_4)\}\,\mathbb{E}\{n(t_2)n^*(t_3)\}. \qquad (13.7)$$

The circular assumption (13.2), which makes the relation function vanish, cancels the second term of this sum, leading to the result:

$$\text{cov}\{S_n(t,\omega), S_n(t',\omega')\} = \gamma_0^2\, S_g(t'-t, \omega-\omega'). \qquad (13.8)$$

As a function of time and frequency differences only, this result expresses the second-order stationarity (or homogeneity) of the spectrogram surface considered as a 2D random field.

Remark. If we had assumed $n(t)$ to be real-valued, there would have been no cancellation of the product term that involves relation functions in the complex case. An explicit calculation shows that whenever $n(t) \in \mathbb{R}$ is a zero-mean white Gaussian noise with correlation $\gamma_n(\tau) = \gamma_0\,\delta(\tau)$, the spectrogram covariance adds to (13.8) an extra term according to:

$$\text{cov}\{S_n(t,\omega), S_n(t',\omega')\} = \gamma_0^2\left[S_g(t'-t,\omega-\omega') + S_g(t'-t,\omega+\omega')\right]. \qquad (13.9)$$

This extra term breaks the 2D stationarity of the spectrogram considered as a random field. What happens, however, is that this breaking of stationarity only concerns the frequency direction, with a significant effect restricted to a band that is centered on the zero frequency and whose width is directly controlled by the (frequency) radius of the reproducing kernel. A similar effect would be observed in the half-domain of positive frequencies if a real-valued white Gaussian noise were made complex by computing its analytic version. While more details can be found in [120], one can keep in mind that the complex white Gaussian noise idealization is a reasonably good model for real-valued or analytic situations, provided that the analysis is performed "not too close" to the time axis.

Applying the covariance (13.8) to the case where $(t,\omega) = (t',\omega')$, it follows that the variance is constant and reads:

$$\text{var}\{S_n(t,\omega)\} = \gamma_0^2. \qquad (13.10)$$

Therefore, although the average spectrogram is flat, any realization is subject to fluctuations whose mean square deviation is of the order of the mean value (which is reminiscent of the statistics of a pure periodogram in classic spectrum analysis).

The covariance formula (13.8) has been written here for a Gaussian spectrogram, but it is more general and holds true for any unit energy window $h(t)$ in place of the circular Gaussian window $g(t)$. Regardless of the window, its interpretation is that:

> The time-frequency correlation between any two spectrogram values is directly controlled by the (squared) reproducing kernel of the analysis.

This gives a statistical counterpart to the geometrical interpretation of redundancy that was attached to the reproducing kernel in Chapter 6.

Turning back to (13.8), the specific use of $g(t)$ is, however, of special interest since we end up with the Gaussian kernel expression

$$\text{cov}\,\{S_n(t,\omega), S_n(t',\omega')\} = \gamma_0^2 \,\exp\left\{-\frac{1}{2}d^2((t,\omega),(t',\omega'))\right\} \tag{13.11}$$

where

$$d((t,\omega),(t',\omega')) = \sqrt{(t-t')^2 + (\omega-\omega')^2} \tag{13.12}$$

measures the Euclidian distance in the plane between the two considered points.

13.3   Logon Packing

The very simple expression (13.11) allows for a first reading of a spectrogram surface of complex white Gaussian noise, such as those plotted in Figure 13.1: in a first approximation, they are made of a random distribution of *Gaussian bumps*, whose time-frequency extension is that of the (squared) reproducing kernel. If we remember that the reproducing kernel is itself the STFT of the circular Gaussian window, it becomes natural to think of a *constructive* model in which a realization of complex white Gaussian noise would be of the form

$$n(t) = \sum_m n_m\, g(t - t_m)\, \exp\{i(\omega_m t + \varphi_m)\}, \tag{13.13}$$

where $\{(t_m, \omega_m); m \in \mathbb{Z}\}$ stands for the time-frequency centers of the logons $g(t)$ in the plane, n_m for suitable weights, and φ_m for possible phase terms. It follows from this expansion that the associated spectrogram would read

$$S_n(t,\omega) = \left| \sum_m n_m\, F_g(\omega - \omega_m, t - t_m)\, \exp\{i\varphi_m\} \right|^2. \tag{13.14}$$

Such an empirical model contains three possible causes of randomness, namely the weights, the locations of the logon centers, and the phases. In order to give sense to a "mean model," we can first assume that the weights are independent and identically distributed so as to reflect stationarity in both time and frequency. This amounts to writing

$$\mathbb{E}\{n_m n_{m''}^*\} = C\,\delta_{mm'}, \tag{13.15}$$

where C is some constant, leading to the average behavior

$$\mathbb{E}\{S_n(t,\omega)\} = C \sum_m S_g(t - t_m, \omega - \omega_m). \tag{13.16}$$

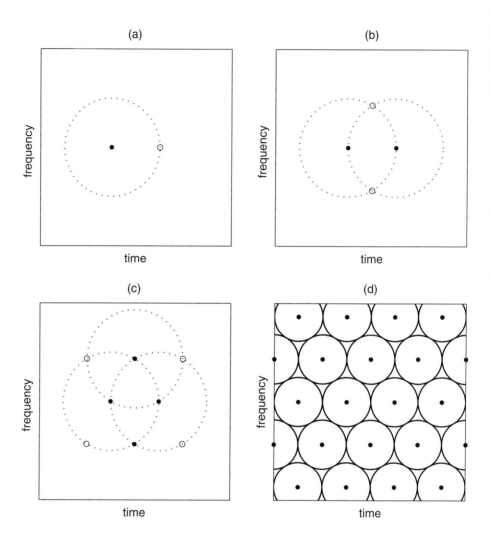

Figure 13.4 White Gaussian noise in the time-frequency plane 3. This figure presents a graphical heuristics for the mean model controlling the time-frequency locations of logon centers. Given a logon centered at some time-frequency location (black dot in (a)), the circular structure of the covariance (13.11) suggests that (on average) a neighboring, independent logon is likely to show up anywhere on the limiting circle (dotted line) from which the covariance can be considered as negligible. Picking up at random such a point (circle), the same argument applies to the next neighbors. Intersecting the two circles (b) leads to logon centers located on the vertices of equilateral triangles and, proceeding further the same way (c), the average distribution of logon centers is expected to form a regular triangular lattice, with effective supports of the corresponding spectrograms that are maximally packed (d). 2017 IEEE. Adapted, with permission, from [121].

If we compare this expression with (13.5), the mean model will be approximately correct if we make the identification $C = \gamma_0$ and if, on the average, summing the Gaussian bumps results in a surface that is almost flat. To this end, we are led to consider a *regular lattice* structure for the locations of the logon centers so as to guarantee the time-frequency homogeneity required by stationarity [121].

A natural choice is to choose a *regular triangular lattice* as the mean model for the logon centers.

The rationale for such a choice can be qualitatively justified for at least two complementary reasons:

1. The first reason comes from a *geometrical* argument. Indeed, we have seen that spectrograms of individual logons – which happen to coincide with the (squared) reproducing kernel of the analysis – are known to concentrate most of their energy in time-frequency domains taking the form of circular disks, with the consequence that organizing them on a regular triangular lattice corresponds to *maximum packing* [122].

2. The second reason comes from a *statistical* argument. Given a logon centered at some time-frequency location, the circular structure of the covariance function (13.11) suggests that (in the mean) a neighboring, independent logon is likely to show up anywhere on the limiting circle from which the covariance can be considered as negligible. Picking up any point on the limiting circle at random, the same argument applies to the next neighbors. Intersecting the two circles results in logon time-frequency centers being located on the vertices of equilateral triangles at the points where the circles intersect; then, if this process is continued in the same manner, the average distribution of logon centers is expected to form a regular triangular lattice.

Both arguments can be merged by resorting to the graphical interpretation given in Figure 6.1, with the optimally packed circles having their centers spaced apart by a distance corresponding to roughly half the radius of the (squared) reproducing kernel, as it can be measured by the level curve at half the maximum amplitude of the Gaussian. This is illustrated in Figure 13.4.

14 More on Maxima

The example of Figure 13.2 stressed the importance of local extrema in the structure of spectrograms of complex white Gaussian noise. It seems worth having a deeper understanding of how such characteristic points are distributed, so as to base further processing on this knowledge. In this chapter, we will consider maxima in closer detail, while Chapter 15 will be devoted to zeros.

14.1 A Randomized Lattice Model

The mean model that was proposed in the previous chapter for the locations of the logon centers takes the form of a regular triangular lattice that is naturally *deterministic*. For the picture to be more complete and more realistic, one has to account for *random fluctuations* around this mean model. As we focus in for the first time on the locations of the logon centers (which will be assumed to coincide with local maxima), the simplest model we can propose for taking into account fluctuations is that of a perturbation of the regular lattice, with random time and frequency shifts. For the sake of simplicity, those shifts will be assumed to be independent and identically distributed, according to a Gaussian distribution [121].

> Within this picture, local maxima can be viewed as a 2D point process in the time-frequency plane.

The simplest model for a planar point process is the *Poisson point process*. In its homogeneous form, such a process corresponds to point locations that are totally independent of each other and drawn under the control of only one parameter, the density λ. This situation, referred to as "Complete Spatial Randomness" [123, 124], is unlikely to be a good model for spectrogram local maxima because of the constraints imposed by the reproducing kernel (see Chapter 6). It can, however, serve as a reference model to which actual data can be compared.

The intrinsic independence that defines a Poisson model can be tested via the distribution of distances between nearest neighbors [123]. Indeed, if we randomly pick a point (t, ω) in the time-frequency plane and assume that it is drawn from a Poisson process, the distribution of the distance to its nearest neighbor can be determined by considering a circle of radius d centered in (t, ω), and by looking for the probability that no other

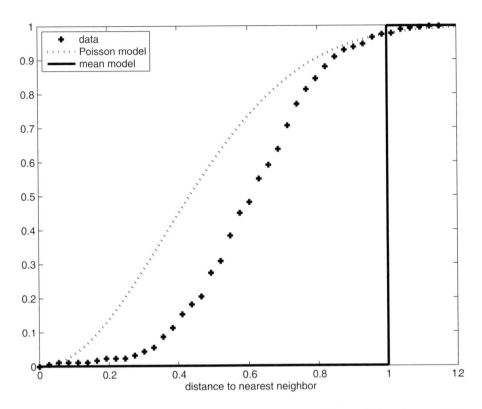

Figure 14.1 Spectrogram extrema are not Poisson distributed. This figure displays the cumulative distribution function of nearest-neighbor distance for actual spectrogram local maxima in the case of complex white Gaussian noise (crosses). It also plots what would be the result for the deterministic mean model in which all maxima are vertices of a regular triangular lattice (solid line), as well as for an equivalent Poisson process with the same density (dotted line). All distances are measured with respect to the edge length of the mean model, chosen as unity.

point is included within this circle, which exactly corresponds to the probability that the distance D to the nearest neighbor is at least d. Given the density λ, this probability reads

$$\mathbb{P}\{D > d\} = \exp\{-\lambda \times \pi d^2\}, \tag{14.1}$$

from which it follows that the complete spatial randomness would imply, for the distance to nearest neighbor, a cumulative distribution function given by:

$$\mathbb{P}\{D \le d\} = 1 - \exp\{-\lambda \pi d^2\}. \tag{14.2}$$

When looking at actual data (see Figure 14.1), it is clear that, as expected, the Poisson model is ruled out, calling for an alternative model that would better describe the observed behavior. To this end, one can build upon an approach outlined in [125], which closely parallels the one which is classically followed in the Poisson case.

More precisely, the method proceeds as sketched in Figure 14.2. Within the mean model, each logon is centered at a point belonging to a regular triangular lattice, meaning

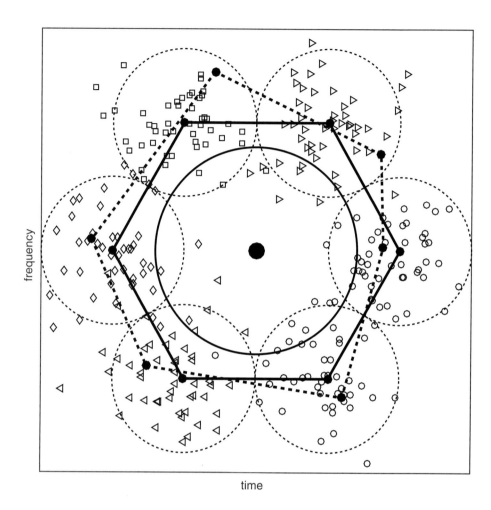

time

Figure 14.2 Randomized lattice model 1. In a regular hexagonal lattice, each node chosen as a reference (big dot) is surrounded by six neighbors that are all at the exact same distance, thus forming a perfect hexagon (solid lines). When randomizing this lattice with a moderate level of Gaussian noise affecting each node, this hexagon is disturbed and transformed into an irregular polygon (dashed lines). Since each of the shifted nodes is assumed to be drawn from a 2D Gaussian distribution (scattered symbols, with dotted circles figuring dispersion), the probability distribution function of the nearest-neighbor distance to some other node can be deduced from the probability that one realization falls within a given distance (solid circle). The overall distribution follows by assuming individual fluctuations of all nodes to be independent.

that it has six nearest neighbors located on the vertices of a hexagon surrounding the center.

Given such a fixed point arbitrarily chosen at the origin of the time-frequency plane, the first step requires the evaluation of the probability of finding one of its six nearest neighbors within a given distance. To do so, we can first evaluate the probability that *one* of its neighbors lies at a distance at least d from the origin. This simply reads

$$P_1(d) = \mathbb{P}\{D > d\} = 1 - \iint_\Omega p(t, \omega) \, dt \, d\omega, \qquad (14.3)$$

where Ω stands for the disk of radius d centered at the origin of the plane and

$$p(t, \omega) = \frac{1}{2\pi\sigma^2} \exp\left\{-\frac{1}{2\sigma^2}\left[(t - m)^2 + \omega^2\right]\right\} \qquad (14.4)$$

if the Gaussian fluctuations – around the chosen node of coordinates $(m, 0)$ – are assumed to be of variance σ^2 in each direction. A change of variables to polar coordinates leads to:

$$P_1(d) = 1 - \int_0^d F(r; m, \sigma^2) dr, \qquad (14.5)$$

with

$$F(r; m, \sigma^2) = \frac{r}{\sigma^2} \exp\left\{-\frac{1}{2\sigma^2}\left(r^2 + m^2\right)\right\} I_0\left(\frac{rm}{\sigma^2}\right), \qquad (14.6)$$

and

$$I_0(x) = \frac{1}{2\pi} \int_0^{2\pi} \exp\{x \cos\theta\} d\theta \qquad (14.7)$$

((14.7) being the modified Bessel function of first kind [126]).

If we further assume that the variance is small enough to ensure that the nearest neighbor results from fluctuations around the only vertices of the first hexagon around the point of reference, the total probability $P_6(d)$ that no neighbor can be found at a distance of a least d from any given point of the perturbed lattice can be approximated by $P_6(d) = (P_1(d))^6$, thus leading to the final result:

$$\mathbb{P}(D \leq d) = 1 - \left(1 - \int_0^d F(r; m, \sigma^2) dr\right)^6. \qquad (14.8)$$

For a given internode distance in the mean model (that we can set to $m = 1$ for the sake of simplicity), the cumulative distribution function (14.8) that is obtained in this manner only depends on one parameter, namely the variance σ^2 (or the mean-square deviation σ) of the fluctuations of the logon centers locations. In the limit $\sigma \to 0$, the randomized lattice tends to be "frozen" in the configuration of the "crystal-like" triangular lattice of the mean model. In terms of the cumulative distribution function of distances to the nearest neighbor, this leads to a degenerate case since all such distances tend to be equal (with value $m = 1$), resulting in a step function. When σ grows, the disorder increases and the randomized lattice has vertices distributed in a more and more random manner, thus ultimately resembling a Poisson process. For intermediate values of σ, neither too small nor too large, the cumulative distribution function evolves continuously from the step function to the Poisson curve.

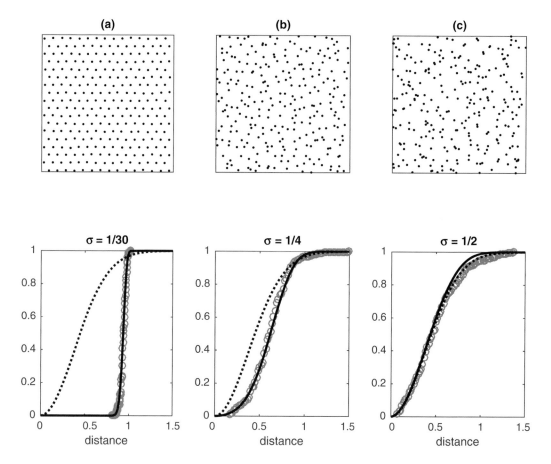

Figure 14.3 Randomized lattice model 2. The three top diagrams from (a) to (c) display sample realizations of the randomized lattice model with $m = 1$, and three different values of σ ranging from $1/30$ (almost no disorder) to $1/2$ (fully developed disorder). The three bottom diagrams present the corresponding cumulative distribution functions of the nearest-neighbor distance for the actual data (gray circles), the equivalent Poisson process with the same density (dotted lines), and the model prediction (solid lines).

This is illustrated in Figure 14.3, with three simulations ranging from almost no disorder ($\sigma = 1/30$) to fully developed disorder ($\sigma = 1/2$), with $\sigma = 1/4$ as an intermediate step. The scatter plots of the corresponding planar point processes are complemented by the corresponding cumulative distribution functions of the nearest-neighbor distance. All diagrams display the actual data, together with the model prediction given by (14.8) and the equivalent Poisson reference for the same density.

To check the relevance of the proposed model for actual spectrogram extrema, we come back to the case of Figure 14.1 and look for the best fit when using the model (14.8). This is reported in Figure 14.4, which shows a fairly good agreement for the best fit choice $\sigma \approx 1/3.3$. This is further supported by Figure 14.5, which compares various realizations of the randomized lattice models to an actual distribution of spectrogram

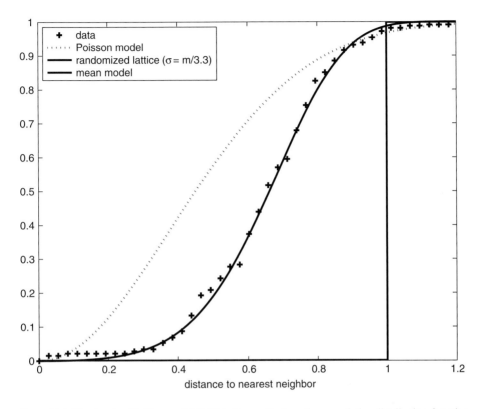

Figure 14.4 Randomized lattice model 3. This figure displays the cumulative distribution function of nearest-neighbor distance for actual spectrogram local maxima in the case of complex white Gaussian noise (crosses). It also plots what would be the result for the deterministic mean model, in which all maxima are vertices of a regular triangular lattice (full line), the distribution for an equivalent Poisson process with the same density (dotted line), as well as the best fit from the randomized model (14.8) (dashed-dotted line). 2017 IEEE. Adapted, with permission, from [121].

maxima. The differences between the point patterns are almost indistinguishable to the naked eye.

14.2   Ordinates and Maxima Distributions

The STFT being a linear transform, it preserves Gaussianity [42]. In particular, the STFT values of a complex white Gaussian noise are distributed according to a Gaussian law. The corresponding spectrogram – i.e., the squared magnitude of the STFT – is composed of two terms, namely

$$S_n(t, \omega) = (\text{Re}\{F_n(t, \omega)\})^2 + (\text{Im}\{F_n(t, \omega)\})^2 . \qquad (14.9)$$

For a given time-frequency location (t, ω), each of those terms therefore appears as a squared Gaussian variable. When standardized, and because of the assumed

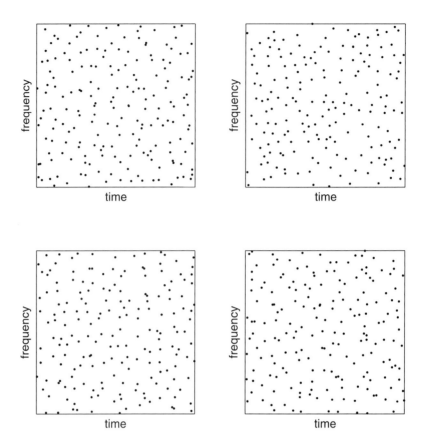

Figure 14.5 Randomized lattice model 4. This figure displays three sample realizations of the randomized lattice model (14.8) with $m = 1$ and $\sigma = 1/3.3$ as in Figure 14.4, as well as one actual realization of spectrogram maxima with the same density. As the art shows, it can be rather difficult to distinguish the samples from the real thing. © 2017 IEEE. Adapted, with permission, from [121].

de-correlation of the analyzed complex white Gaussian noise, their sum is expected to be distributed according to a *chi-square law* with 2 degrees of freedom, i.e., an *exponential distribution*.

Remark. Chi-square and exponential distributions both enter the general class of *Gamma distributions* defined by the probability distribution function:

$$p_\Gamma(u; v, \theta) = \frac{u^{v-1} \exp\{-u/\theta\}}{\theta^v \Gamma(v)} \mathbf{1}_{[0,+\infty)}(u), \tag{14.10}$$

where $v > 0$ and $\theta > 0$ are the so-called *shape* and *scale* parameters, respectively. In the case where $v = 1$, the Gamma distribution $p_\Gamma(u; 1, \theta)$ reduces to an exponential distribution. The ordinary chi-square distribution is also a special case of the Gamma distribution, with the one-parameter probability distribution function $p_{\chi^2}(u; k) = p_\Gamma(u; k/2, 2)$, in which k stands for the *number of degrees of freedom*, i.e., the number of independent

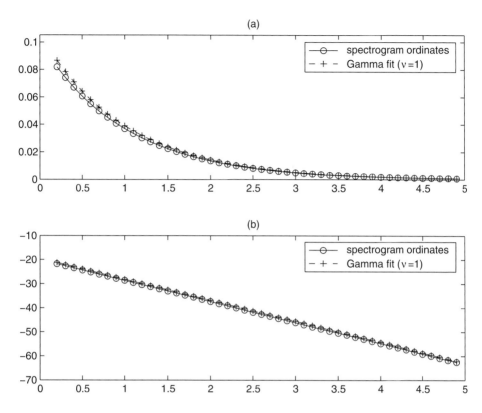

Figure 14.6 Distribution of spectrogram ordinates. In the case of complex white Gaussian noise, spectrogram ordinates follow an exponential distribution. This is proven here by a Gamma fit on the probability distribution function – displayed on both a linear (a) and a logarithmic (b) scale – resulting in a shape parameter $v = 1.003$, i.e., almost 1, as is expected in the case of an exponential distribution. Here, the probability distribution function has been estimated on spectrogram data renormalized by the mean, and statistics have been evaluated on 1,000 realizations of complex white Gaussian noise sequences with 256 samples each.

squared Gaussian variables that enter the sum. It follows that, in the case where only two degrees of freedom are involved, $p_{\chi^2}(u; 2)$ also reduces to the probability distribution function of an exponential distribution.

According to this remark, then, fitting a Gamma model to data is a way to check for the relevance of an exponential model, if any. This is what is done in Figure 14.6, where the estimated value $v \approx 1$ of the shape parameter clearly supports the claim that:

> Spectrogram ordinates of complex white Gaussian noise are distributed according to an exponential law.

Turning to the distribution of the spectrogram maxima calls for the theory of extreme values [127, 128]. Without entering into the details of this theory, one of its key results is that the probability distribution function of extrema can only take on three very specific forms. More precisely, the most general form for the probability distribution function of the maximum of a set of independent and identically distributed random variables is given by the *Generalized Extreme Value* (GEV) parameterization:

$$p_{GEV}(u; k, \sigma, \mu) = \frac{1}{\sigma}(1 + kv)^{-(1+1/k)} \exp\left\{-(1 + kv)^{-1/k}\right\}, \qquad (14.11)$$

with $v = (u - \mu)/\sigma$ a standardized version of u, and where the parameters $\mu \in \mathbb{R}$ and $\sigma > 0$ stand for *location* and *scale*, respectively, while the *shape* parameter $k \in \mathbb{R}$ permits a classification in three different *types* of distributions.

Whenever $k > 0$ or $k < 0$, the distributions are classified as "type II" or "type III" (with the corresponding distributions referred to as *Fréchet* or *Weibull*), respectively. The case where $k = 0$ ("type I") has to be considered as a limit, according to

$$p_G(u; \sigma, \mu) = \lim_{k \to 0} p_{GEV}(u; k, \sigma, \mu) \qquad (14.12)$$

$$= \frac{1}{\sigma} \exp\left\{-[(u - \mu)/\sigma + \exp\{-(u - \mu)/\sigma)\}]\right\}, \qquad (14.13)$$

and the corresponding distribution is referred to as *Gumbel*. Let us remark that, in this last expression, the mean value $\bar{\mu}$ of the distribution is related to the location and scale parameters according to $\bar{\mu} = \mu + \gamma\sigma$, with $\gamma \approx 0.5772$ being the Euler-Mascheroni constant.

For those expressions to make sense, the initial distribution of the considered random variables (whose distribution of the maximum must be determined) has to be in the domain of attraction of the laws described by (14.11). It turns out [128] that this is the case for the exponential distribution with respect to the Gumbel distribution (14.13). It is therefore expected that, when considering a spectrogram of complex white Gaussian noise as made of identically distributed random variables that all follow the same exponential law, its *global* maximum will follow a Gumbel law.

As illustrated in Figure 14.7 (top diagram), extensive numerical simulations support this claim. Fitting a Generalized Extreme Value distribution to actual data yields a shape parameter $k = -0.02$, which is fairly close to the value $k = 0$ that is characteristic of a Gumbel distribution.

A similar analysis on the set of all *local* maxima (bottom diagram) yields a distribution in a reasonably good agreement with a Gumbel model (we obtain $k = 0.08$ in this case).

> The global maximum and the local maxima of the spectrogram of complex white Gaussian noise are distributed according to a Gumbel distribution.

Note that the scale parameter σ is similar for global and local maxima, with respective values 1.2 and 1.1.

Remark. While the shape parameter k has an absolute meaning, both the location and scale parameters (μ and σ, respectively) have only a relative meaning that depends on the power spectrum density level of the analyzed complex white Gaussian noise and,

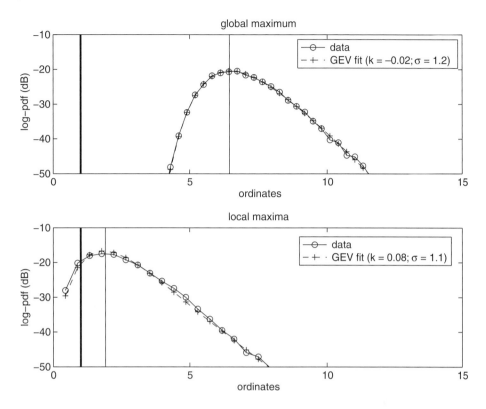

Figure 14.7 Distribution of maxima for the spectrogram of complex white Gaussian noise. For both the global maximum (top diagram) and the local maxima (bottom diagram), actual data have a probability distribution function that can be well fitted by a Generalized Extreme Value distribution with a shape parameter k that is almost zero, i.e., a Gumbel distribution. The probability distribution functions are plotted in logarithmic units (dB), the thin vertical lines indicate the location parameters μ of the distributions, and the thick vertical lines indicate the theoretical mean value of the spectrogram, i.e., the power spectrum density $\gamma_0 = 1$.

hence, on the effective conditions of the numerical simulations. In the case of Figure 14.7, all data sequences have been generated such that the mean power spectrum level is set to unity. Each sequence consists of 256 data samples, with 1,000 independent realizations for local maxima and 25,000 for the global maximum (so as to get statistics with comparable data volumes.) It has already been mentioned that, in the Generalized Extreme Value distribution fits, the *absolute* variability is of the same order of magnitude for both the global maximum and the local maxima ($\sigma = 1.2$ and $\sigma = 1.1$, respectively). However, when we consider the location parameter of the distribution (its "mode"), we get $\mu = 6.45$ and $\mu = 1.90$, respectively, with the result that local maxima experience a much higher *relative* variability than the global maximum, with a ratio $\sigma/\mu = 0.59$ to be compared with the value $\sigma/\mu = 0.18$ for the global maximum distribution.

One interesting point of the semi-empirical study of the local maxima distribution is that it allows us to complete the proposed model (13.14) by imposing a distribution

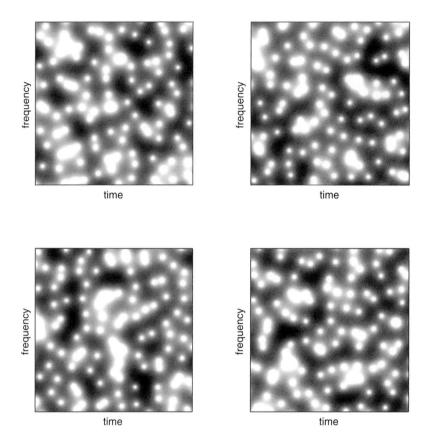

Figure 14.8 Synthetic spectrograms. This figure compares an actual spectrogram of complex white Gaussian noise with three sample realizations of the (random phases) model (13.14), which is based on a randomized lattice structure for the logon locations and a Gumbel distribution for drawing their weights, with the free parameters fixed by the best fits reported in Figures 14.4 and 14.7. As the art shows, it is rather difficult to distinguish the samples from the real thing. © 2017 IEEE. Adapted, with permission, from [121].

of the weights n_{mk} that is controlled by the assumed Gumbel distribution of the local maxima (together with phase values that have been chosen as uniformly distributed over each local period so as to maximally "unlock" the constitutive logons). By adjusting the free parameters (variance of the fluctuations in the randomized lattice model and scale/location in the Gumbel model) to the best fits reported in Figures 14.4 and 14.7, we obtain synthetic spectrograms such as those displayed in Figure 14.8.

14.3   Voronoi

At the end of Chapter 9 we mentioned the connection that exists, in the "many logons" case, between basins of attraction and Voronoi tessellations constructed on local maxima. Considering both the experimental evidence of "time-frequency patches" reported

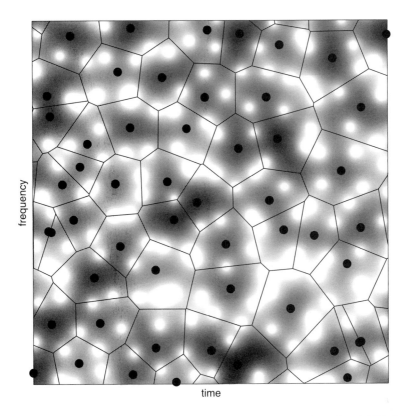

frequency

time

Figure 14.9 Spectrogram and Voronoi tessellation based on local maxima 1. This figure displays an example of a spectrogram of complex white Gaussian noise, with local maxima highlighted as large dots and the attached Voronoi tessellation superimposed as thin lines.

at the beginning of Chapter 13 and the simplified model developed thereafter, complex white Gaussian noise naturally offers an instance of such a "many logons" situation.

Figure 14.9 shows an example of a Voronoi tessellation constructed on spectrogram local maxima in the case of complex white Gaussian noise. By construction, it presents a structure made of polygonal convex cells of varying sizes and shapes. Despite this randomness, it presents a statistical homogeneity that reflects the spatial (time-frequency) homogeneity of the maxima point process that stems itself from the time stationarity of the analyzed process.

If the maxima point process were to be Poisson, a whole literature about the statistics of the corresponding Voronoi cells would be at hand (see, e.g., [129] or [130]). Unfortunately, however, this is not the case, and, like many other authors have (even in the Poisson case [131]), we must resort to numerical simulations. Two findings are reported in Figure 14.10, obtained with a set of 1,000 independent realizations of complex white Gaussian noise sequences of 1,024 data points each. The first finding (diagram (a)) shows more precisely the variability of the Voronoi cells in terms of *size*, with a probability distribution function that can be reasonably well fitted by a Gaussian distribution.

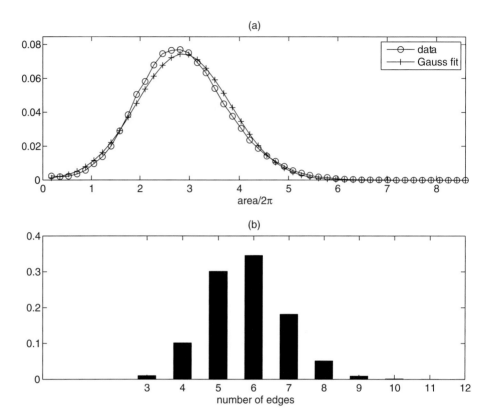

Figure 14.10 Spectrogram and Voronoi tessellation based on local maxima 2. In the case of complex white Gaussian noise, the area of the Voronoi cells attached to local spectrogram maxima is a random variable with a distribution that is approximately normal (diagram (a)). The empirical distribution of the number of edges of those cells reveals that they are mostly hexagonal (diagram (b)).

The second finding (diagram (b)) concerns the *shape* of the cells by evaluating the distribution of the number of their edges. What turns out is that the empirical mean of this number (≈ 5.98) is very close to 6, which is also the mode. In brief,

> Most of the Voronoi cells constructed on the local spectrogram maxima of complex white Gaussian noise are hexagonal.

This is, of course, consistent with the mean model based on a regular triangular lattice. It supports the randomized lattice model as an intermediate between the fully ordered mean model and the maximally disordered Poisson model.

Remark. In this case, it is worth mentioning a work by Valerio Lucarini [132] that is of special relevance for this situation. Indeed, among various models of non-Poisson point processes, one of those that this author investigated (from a merely numerical perspective) was precisely that of a regular triangular grid disturbed by random Gaussian shifts

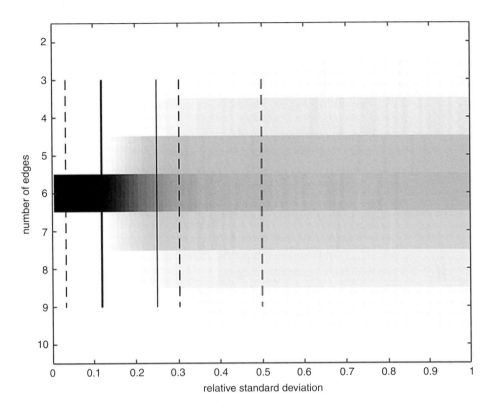

Figure 14.11 Randomized lattice model and phase transition. This figure displays the empirical distribution of the number of edges of Voronoi cells in the case of the randomized lattice model, as a function of the relative standard deviation of the vertices fluctuations with respect to the triangular grid spacing. The thick vertical line indicates the threshold mentioned in [132] for the phase transition that breaks the exact hexagonal symmetry. Dashed vertical lines correspond to the values reported in the model diagrams of Figure 14.3, and the thin line corresponds to the value estimated on real data in Figure 14.4.

of its vertices, i.e., the randomized lattice model considered in this chapter. Lucarini's analysis shows that the hexagonal tessellation ("honeycomb" structure) is "typical" in the sense that it remains stable under small disturbances of its vertices. For small amounts of random shifts (i.e., below a threshold value of the shift standard deviation, which has been found to be about 8.3 times smaller than the grid spacing), hexagons are the only structure to be observed. For larger values of the shifts, hexagons remain the most likely structure, in addition to possessing characteristics (such as area and shape) that quickly boil down to those of a Poisson process. This *phase transition* effect is illustrated in Figure 14.11, which shows the distribution of the number of edges of each cell as a function of the relative standard deviation, which acts as a control parameter: when some threshold is passed, the support of this distribution is changed.

To conclude this chapter on spectrogram maxima, one can mention that using such characteristic points as a relevant simplification of a time-frequency surface is quite natural and has been effectively used in several situations. Let us mention only two of them. The first one is in the audio domain, where reducing a whole spectrogram to a scatter plot consisting of the time-frequency locations of its "significant" local maxima provides a *fingerprint* that can be used as a template for recognition. This rationale is at the heart of actual industrial audio recognition systems [133]. The second one is in physics (namely in experimental astrophysics aimed at the detection of gravitational waves), where interferometric observations recorded at two independent detectors are first transformed into a time-frequency spectrogram-like representation [134]. Local maxima in such diagrams are then retained, provided they correspond to a "significant" excess of energy as compared to what would happen in a noise-only situation, and detection is actually achieved by comparing the fingerprints of the two detectors (we will come back to this example in Chapter 16).

In both cases, one key point is to assert the "significance" of local maxima as a signature of some signal component differing from fluctuations attached to noise. It is believed that a better knowledge of the noise-only situation, as it has been considered in this chapter, may help in the decision. In practice, scatter plots made of perfectly localized points may prove much too precise for a meaningful comparison as soon as observations are disturbed, e.g., by additive noise. This calls for ways of *making* the *approach more robust*. One possibility – which makes complete sense if we recall the finite extent of the reproducing kernel of the spectrogram analysis and of the impossibility of any perfectly localized time-frequency signature – is to extend the scatter plot to a collection of *domains* surrounding the time-frequency points to which they are anchored. Going on with local maxima, this can be done with the attached Voronoi cells. However, it turns out that a complementary perspective can also be envisioned, in which the local signal energy contributions are rather identified via domains which exist *in between zeros*. This prompts us to have a closer look at those other extrema that are zeros: this is the topic of the next chapter.

15   More on Zeros

If we return to the example of Figure 14.9, we can complement the description we made of the spectrogram surface in terms of local maxima and Voronoi tessellation by the distribution of its zeros. This is what is done explicitly in Figure 15.1.

As can be seen from this figure, and apart from the fact that zeros are more numerous than local maxima, it is striking that they present some networked structure, tending to organize on (or at the vicinity of) the edges of the Voronoi cells. This feature, which had been previously justified by interference arguments from an abstract point of view (and supported by a toy simulation only; see Figure 9.4), is effectively illustrated here in an actual and meaningful situation.

Focusing further on zeros only, another remarkable feature is that they appear as neither too close nor too spaced apart. In order to determine precisely how close they can be, we can first adopt a phenomenological point of view and make use of the "Fry diagram" [135], which consists of plotting the difference vectors $\mathbf{x}_i - \mathbf{x}_j$ for all pairs of time-frequency points $\mathbf{x}_i$ of coordinates t_i and ω_i. Among other characteristics (such as clustering), a Fry diagram permits us to visualize a possible *short-range repulsion* between points by preserving an almost clear area in the vicinity of the origin (this would be a perfectly clear area in the case of a hard-disk model attaching some nonzero area domain to each point). Indeed, Figure 15.2 reveals such a structure (see the left diagram), while the density estimated from an average of such diagrams over 1,000 independent realizations (right diagram) gives a first idea of the distribution of the distance between neighboring points. Besides short-range repulsion, this clearly hints at the rotation invariance (or isotropy) of the distribution.

This chapter will go further and consider spectrogram zeros as characteristic points whose typical geometry will be explored in the case of complex white Gaussian noise, with an application to filtering out chirplike signals from noise. As for most of this book, emphasis is made on basic ideas, general principles, and interpretations. Interested readers are referred to [120] for a more rigorous approach to many of the questions addressed in this chapter.

15.1   Factorizations

In order to fully understand the importance of spectrogram zeros, let us start again from the Bargmann factorization of the STFT, as stated in (12.2) and (12.5). Recalling

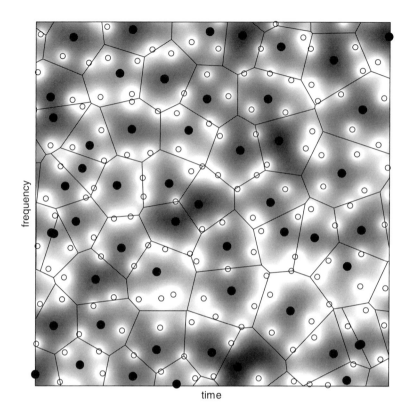

Figure 15.1 Spectrogram, Voronoi tessellation based on local maxima, and zeros. This figure replicates the diagram of Figure 14.9 with zeros highlighted by circles, showing that they are mostly located on (or at the vicinity of) the edges of the Voronoi cells constructed on local maxima.

that the short-time Gaussian window $g(t)$ is of unit energy, it immediately follows that

$$|\mathcal{F}_x(z)| \leq \|x\| \exp\left\{\frac{1}{4}|z|^2\right\},$$ (15.1)

hence

$$\limsup_{|z|\to\infty} \frac{\log\log|\mathcal{F}_x(z)|}{\log|z|} = 2,$$ (15.2)

i.e., that $\mathcal{F}_x(z)$ is an *entire function* of order 2 [136]. As a result, it permits a *Weierstrass-Hadamard factorization* of the form

$$\mathcal{F}_x(z) = z^m \exp\{P_2(z)\} \prod_n \left(1 - \frac{z}{z_n}\right) \exp\left\{\frac{z}{z_n} + \frac{1}{2}\left(\frac{z}{z_n}\right)^2\right\}.$$ (15.3)

In this expression, the variables $z_n = \omega_n + it_n$ stand for the (possibly infinitely many) zeros of the Bargmann transform, $m \in \mathbb{N}$ corresponds to a possible m-fold zero at the origin of the plane, and $P_2(z) = C_0 + C_1 z + C_2 z^2$ is a polynomial of order at

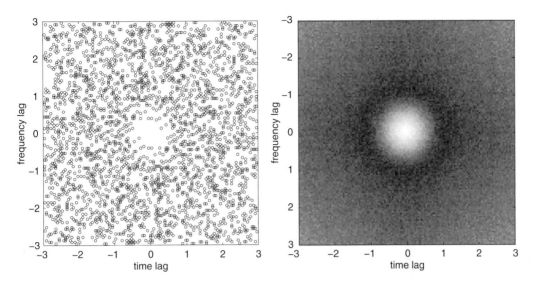

Figure 15.2 Fry diagram of spectrogram zeros in the case of complex white Gaussian noise. Plotting the difference vectors $\mathbf{x}_i - \mathbf{x}_j$ for all pairs of time-frequency points $\mathbf{x}_i$ of coordinates t_i and ω_i reveals an almost clear area around the origin of the plane, which is the signature of a short-range repulsive interaction between the points. The isotropic density of an average of such diagrams over 1,000 different, independent realizations (right diagram) gives an estimate of the distribution of the distance of each point to its neighbors.

most 2, with complex-valued coefficients. As commented in Appendix B of [137], this polynomial allows for simple geometrical transformations such as normalization (C_0), translation/rotation (C_1) and squeezing (C_2). In summary, since the zeros of the Bargmann transform coincide with those of the Gaussian STFT and, hence, of the associated spectrogram, we can consider that

> The structure of a Gaussian STFT/spectrogram is essentially determined by the zeros of the transform.

In the simple case of the Gabor logon, we know from (6.9) that the STFT reads $F_g(t, \omega) = \exp\{-(t^2 + \omega^2)/4\}$. Thus, it has no zero and we get $\mathcal{F}_g(z) = 1$. We can then turn to the more general family of *Hermite functions*, of which the Gabor logon is the first member, and which we previously considered an interesting example (see Figure 10.7). For $k = 0, 1, \ldots$, this family is expressed as

$$h_k(t) = (2\pi)^{-1/4} c_k H_k(t) \exp\{-t^2/2\}, \tag{15.4}$$

with $H_n(\alpha)$ the *Hermite polynomials* defined by

$$H_k(t) = (-1)^k \exp\{\alpha^2\} \frac{d^k}{d\alpha^k} \exp\{-t^2\}, \tag{15.5}$$

and $c_k = (2^{k-1/2} k!)^{-1/2}$.

We obtain in this case [19, 95]

$$\mathcal{F}_{h_k}(z) = \frac{(-i)^k}{2^{\frac{k}{2}} \sqrt{k!}} z^k, \tag{15.6}$$

with a corresponding spectrogram that is overall strictly positive, except at the origin where it presents a kth-order zero.

Beyond these oversimplified examples, which are strongly structured and almost completely controlled by the nonvanishing part of the factorization (15.3), we can expect that disordered situations such as those encountered for complex white Gaussian noise will involve many zeros that will convey most of the information. Indeed, recalling that the family of Hermite functions $\{h_k(t); k = 0, 1, \ldots\}$ constitutes an orthonormal system, the linearity of the Bargmann transform leads to the expansion

$$\mathcal{F}_n(z) = \sum_{k=0}^{\infty} \frac{(-i)^k}{2^{\frac{k}{2}} \sqrt{k!}} \langle n, h_k \rangle z^k \tag{15.7}$$

when applied to complex white Gaussian noise $n(t)$,

Still due to linearity, the inner products $\{\langle n, h_k \rangle; k = 0, 1, \ldots\}$ are *complex Gaussian random variables*. Moreover, they are identically distributed and independent, due to the orthonormality of the expansion system. The assumptions made in (13.1)–(13.3) guarantee that the coefficients $n_k = \langle n, h_k \rangle$ in the expansion (15.7) have a circular Gaussian probability distribution function $p(n) = (1/\gamma_0 \pi) \exp\{-|n|^2/\gamma_0\}$. In other words, they obey:

$$\mathbb{E}\{n_k\} = 0; \tag{15.8}$$

$$\mathbb{E}\{n_k n_{k'}\} = 0; \tag{15.9}$$

$$\mathbb{E}\{n_k n_{k'}^*\} = \gamma_0 \delta_{kk'}, \tag{15.10}$$

where $\delta_{kk'}$ stands for the Kronecker symbol, whose value is 1 when $k = k'$ and 0 elsewhere. Thus, it follows that

$$\mathbb{E}\{\mathcal{F}_n(z)\mathcal{F}_n^*(z')\} = \gamma_0 \exp\left\{\frac{1}{2}zz'^*\right\} \tag{15.11}$$

and, hence, using (12.2) and the usual simplification of notation, that

$$\mathbb{E}\{F_n(z)F_n^*(z')\} = \gamma_0 \exp\left\{-\frac{1}{4}|z - z'|^2 + \frac{i}{2}\mathrm{Im}\{zz'^*\}\right\}. \tag{15.12}$$

Setting $z = \omega + it$, $z' = \omega' + it'$, and $\gamma_0 = 1$, we recognize in (15.12) the reproducing kernel of the STFT for a circular Gaussian window (see (6.4) and (6.8)), whereas (15.11) defines the corresponding reproducing kernel in the Bargmann space of analytic functions. Proceeding further with spectrograms expressed as $S_x(t, \omega) = |F_x(z)|^2$, it is easy to check that

$$\mathrm{cov}\left\{|F_n(z)|^2, |F_n(z')|^2\right\} = \gamma_0^2 \exp\left\{-\frac{1}{2}|z - z'|^2\right\}, \tag{15.13}$$

which is just another way of writing (13.11).

With some changes of variables, the expansion (15.7) defines a so-called *Gaussian analytic function* [138], whose standard form usually reads

$$\psi(z) = \sum_{k=0}^{\infty} \frac{\zeta_k}{\sqrt{k!}} z^k, \tag{15.14}$$

in which $\{\zeta_k; k = 0, 1, \ldots\}$ are independent standard complex-valued Gaussian random variables with probability density function $p(\zeta) = (1/\pi) \exp\{-|\zeta|^2\}$.

> The study of spectrogram zeros happens to coincide, in the case of complex white Gaussian noise, with that of Gaussian analytic functions.

Remark. It should be noted that a similar situation has been considered in quantum mechanics to characterize possible chaotic maps on the basis of the so-called "stellar representation" of zeros in the Husimi distribution [139]. In this case, the corresponding model (15.14) is referred to as CAZP (for "Chaotic Analytic Zeros Points"). In order to have a complete equivalence between (15.7) and (15.14), it is sufficient to normalize the variance of the inner products $\langle n, h_k \rangle$ to unity, and to make the substitution

$$z = \omega + \mathrm{i}t \mapsto \frac{-\mathrm{i}}{\sqrt{2}} z = \frac{t - \mathrm{i}\omega}{\sqrt{2}}, \tag{15.15}$$

which exactly coincides with the convention that is commonly used for z^* in quantum mechanics, with position and momentum in place of time and frequency, respectively (see, e.g., [140] or [137]).

In the rest of this chapter, we will follow the "unorthodox" path of using zeros rather than nonzero values, making use of results on Gaussian analytic functions as established, in particular, in [138].

15.2 Density

We have seen with (13.8) that the stationarity of complex white Gaussian noise carries over to the spectrogram considered as a random field. The distribution of spectrogram zeros is therefore homogeneous all over the time-frequency plane, and the first question we may ask is that of its *density* or, in other words, of how many of them are contained in a given time-frequency domain.

Approach 1: In pure mathematical terms, we can investigate in detail the "flat" CAZP model (15.14) of Gaussian analytic functions. This is what can be found, e.g., in [138], where it is established ("Edelman-Kostlan formula") that the distribution of zeros of a Gaussian analytic function $f(z)$ of covariance $K_f(z, z')$ has for density

$$\rho_f(z) = \frac{1}{4\pi} \Delta \log K_f(z, z) \tag{15.16}$$

with respect to the Lebesgue measure on the complex plane. Applying this result to the current time-frequency setting with the conventions $\omega + it$ for z and making use of (15.11) with $\gamma_0 = 1$, it follows that – in the case of complex white Gaussian noise – the density of spectrogram zeros reduces to $\rho_n(z) = 1/2\pi$, i.e., that

> One zero is expected per 2π area in the time-frequency plane.

Approach 2: Another, more phenomenological, way to approach this question is to follow a reasoning in terms of *number of degrees of freedom*. From a conceptual point of view, the rationale comes from the decomposition of (15.3) which, if we interpret it as a complete representation of the transform, suggests that there are essentially as many free parameters as zeros. As a result, this supports the idea that their number can be viewed as a measure of the number of degrees of freedom of the analyzed signal.

In order to follow up with this interpretation, it is easier to think in a discrete-time setting in which the number of degrees of freedom of a complex white Gaussian noise *sequence* is twice the number N of its samples (two values, one real and one imaginary, per sample, all independent). From a (discrete) time-frequency perspective, these $2N$ degrees of freedom have to be traded for a number M of zeros over the principal time-frequency domain $[1, N] \times [-\pi, +\pi]$. Since a spectrogram zero is characterized by two numbers (its time and frequency coordinates), we must have $M = N$ in order to encode the $2N$ degrees of freedom. In a nutshell,

> A complex white Gaussian noise sequence of N samples has a spectrogram which contains N zeros in the principal time-frequency domain $[1, N] \times [-\pi, +\pi]$.

In order to connect the two approaches, one can consider that, in the case of a sampling rate fixed to unity, the principal time-frequency domain $[1, N] \times [-\pi, +\pi]$ has for "area" $2\pi N$, with the result that N zeros in such a domain is equivalent to a density $1/2\pi$.

This claim about density is fully supported by extensive numerical simulations. For instance, an experiment conducted over 250 independent realizations of complex white Gaussian noise sequences, with $N = 2{,}048$ samples each, typically leads to a relative mean number of zeros in the half domain of positive frequencies (with respect to the actual expected mean $N/2 = 1{,}024$) which has a value of $\bar{n}_z = 1.0082$, with a relative standard deviation of $\bar{\sigma}_z = 4.4 \times 10^{-3}$.

15.3   Pair Correlation Function

The Fry diagram of Figure 15.2 shows a short-range repulsive effect in the distribution of zeros: zeros are significantly less likely to be close than in a Poisson distribution, where all distances – be they large or small – between zeros.

One way of quantifying repulsiveness is to make use of the so-called *pair correlation function* $\rho_f(z_1, z_2)$, whose loose interpretation is that, for two given time-frequency points z_1 and z_2, the quantity $\rho_f(z_1, z_2)\mathrm{d}^2(z_1)\mathrm{d}^2(z_2)$ measures the probability of a "coincidence" within infinitesimal neighborhoods.

When considering spectrogram zeros as defining a point process in the time-frequency plane, the combination of the observed repulsion and of the omnipresence in the analysis of an underlying reproducing kernel suggests turn to *determinantal point processes* that were precisely introduced by Odile Macchi [141] for modeling fermion statistics in quantum optics. In the case of a determinantal point process, the pair correlation function is simply expressed as the determinant of a kernel matrix that fully specifies the process (see, e.g., [142] or [143] for further details). This is the case, for instance, with the *Ginibre ensemble*, which consists of the eigenvalues of Gaussian matrices with independent complex-valued entries. The Ginibre ensemble forms a determinantal point process, with a kernel that simply identifies with the reproducing kernel (15.11), thus leading to a pair correlation function that is isotropic and of the form:

$$\rho_f(r) = 1 - \exp\{-\beta r^2\}, \tag{15.17}$$

with $r = |z_1 - z_2|$ and $\beta > 0$.

In contrast, the pair correlation function of Gaussian analytic functions has been evaluated in [138], leading to the closed form expression that was first given in [144]:

$$\rho_f(r) = \frac{(\cosh^2 v + v^2)\cosh v - 2v\sinh v}{\sinh^3 v} \tag{15.18}$$

with $v = \alpha r^2$ and $\alpha > 0$.

As compared to the Ginibre case, this "Hannay universal function" predicts an overshoot above 1 that is actually observed when running the simulation displayed in Figure 15.3. The presence of this overshoot rules out the possibility of describing the distribution of zeros by a determinantal point process since, if this was the case, we should have $\rho_f(r) \leq 1$ [143].

15.4   Voronoi

Proceeding as for local maxima, zeros can be used as cell centers to get a Voronoi tessellation of the time-frequency plane. This results in a random tiling that covers the plane in a stationary (or homogeneous) way. A typical example is given in Figure 15.4, which shows quite a strong regularity in the geometry of the cells.

A numerical investigation (stemming from 250 independent realizations of complex white Gaussian noise sequences with $N = 2,048$ samples each) ends up with the more precise results that are summarized in Figure 15.5.

Area: It turns out that the probability distribution function of area is remarkably well described by a Gaussian, with a mean value $\mu = 0.999 \times 2\pi$ (i.e., almost 2π) which fully corroborates the claim that one zero is expected per 2π area in the time-frequency plane. The standard deviation is estimated to be $\sigma = 0.165 \times 2\pi$ (i.e., almost $\pi/3$), which means that almost all areas are contained in the interval $[\pi, 3\pi]$.

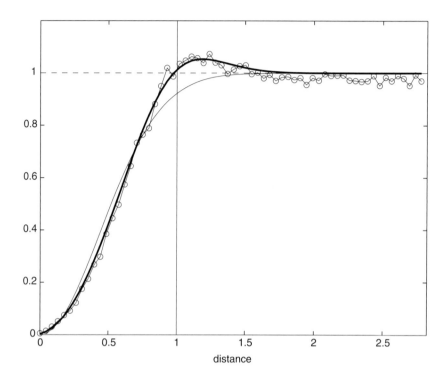

distance

Figure 15.3 Pair correlation function. This figure displays as circles the estimated pair correlation function of spectrogram zeros based on 10,000 independent realizations of 256 samples long sequences of complex white Gaussian noise. The thick and thin lines correspond to fits with the Hannay model (15.18) and the Ginibre model (15.17), respectively.

This situation of Voronoi cells with 2π area on average can be connected to the idea of *gravitational allocation* that is pushed forward and illustrated in [145] and [138]. Given a Gaussian analytic function of the form (15.7), the rationale is to think of

$$U(z) = \log |\mathcal{F}_n(z)| - \frac{1}{4}|z|^2 \qquad (15.19)$$

as a *random potential function*, and to characterize the corresponding gradient curves, which are defined by the evolution equation of a particle:

$$\frac{dZ(\tau)}{d\tau} = -\nabla U(Z(\tau)). \qquad (15.20)$$

Phrased in STFT/spectrogram terms, it directly follows from (15.19) that we have equivalently $U(z) = \log |F_n(z)| = (1/2) \log S_n(t, \omega)$, and (15.20) is of course reminiscent of the *differential reassignment* discussed in Chapter 12 (see (12.12) and (12.14)). The notable difference with the reassignment case is the minus sign that appears in (15.20), which amounts to describing evolutions that converge toward *minima* instead of maxima. The variable Z can therefore be viewed as a particle subject to a gravitational field, falling down from any initial position on the log-spectrogram surface along gradient

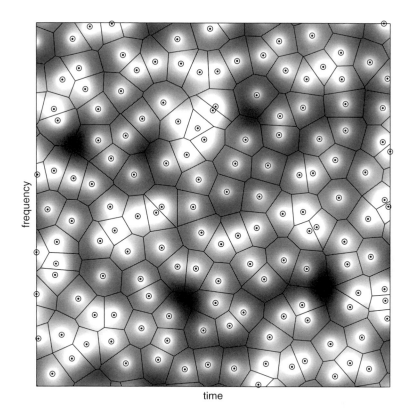

frequency

time

Figure 15.4 Spectrogram and Voronoi tessellation based on zeros 1. This figure considers an example as in Figure 14.9, with zeros highlighted as large dots and the attached Voronoi tessellation superimposed as thin lines.

curves that eventually converge to a zero. Figure 15.6 illustrates such a surface, in the spirit of [145] and [138].

Reversing the perspective, a *basin of attraction* consisting of all initial positions of Z can be attached to each zero such that the evolution governed by (15.20) terminates at that zero. Since each time-frequency point belongs to one and only one such basin, the entire plane is ultimately partitioned in basins that each contain one and only one zero. The remarkable result established in [145] and [138] is that:

> All basins of attraction of zeros are of equal area, namely 2π with our conventions.

Although those basins have no reason to be polygonal (i.e., made of segments of straight lines), it turns out that, in terms of mean area, Voronoi cells give a fairly good approximation of them. Furthermore, if we come back to reassignment, it is remarkable (yet somehow expected) that most of the reassigned energy localizes precisely on the edges of the Voronoi cells centered on zeros, as illustrated in Figure 15.7.

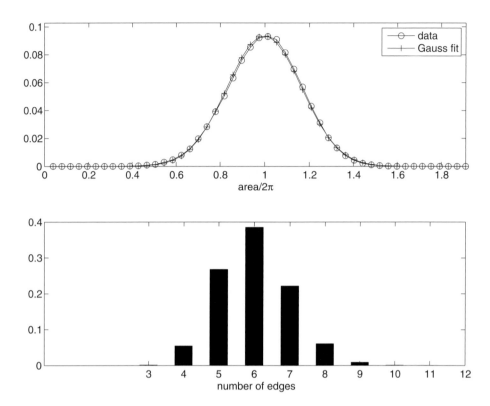

Figure 15.5 Spectrogram and Voronoi tessellation based on zeros 2. In the case of complex white Gaussian noise, the area of the Voronoi cells attached to local zeros is a random variable that is normally distributed, with mean unity and standard deviation close to $1/6$ in 2π units (top diagram). The empirical distribution of the number of edges of those cells reveals that most of them are hexagonal (bottom diagram).

This shows that Voronoi cells based on spectrogram zeros are intimately coupled with the basins induced by gravitational allocation. Alternatively, they can also be seen as a by-product of reassignment.

Shape: Just as with local maxima (cf. Figure 14.10), the Voronoi cells centered on zeros have a range of different numbers of edges, of which the mode is 6. Therefore (again, as with local maxima), hexagonal cells are the most probable configuration.

Voronoi cells based on local maxima or on zeros define two intertwined networks that gain to be jointly considered, as done in Figure 15.8. Beyond the already-mentioned fact that zeros tend to localize along the edges of Voronoi cells attached to local maxima, it appears that local maxima tend to localize at multiple junction points of Voronoi cells attached to zeros. This supports the idea that zeros define a mean honeycomb structure made of hexagons centered on the vertices of the hexagons, which themselves define the larger honeycomb structure attached to local maxima. Within this schematic picture, the area of those local maxima hexagons should be three times that of the zeros' hexagons (see Figure 15.9). This is roughly supported by numerical simulations (see Figure 14.10,

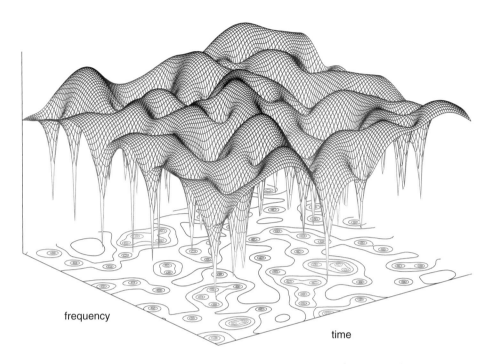

frequency

time

Figure 15.6 Gravitational allocation. This figure displays the log-spectrogram (surface and projection on the time-frequency plane) of a realization of complex white Gaussian noise. Following [145] and [138], this can be interpreted as a potential function governing the falling down of particles (along gradient curves) toward the zeros of the transform.

with an estimated mean area $2.891 \times 2\pi$), and graphically illustrated in Figure 15.8 by means of circles of respective areas 2π and $3 \times 2\pi$.

15.5   Delaunay

The graph formed by the edges of the Voronoi cells based on a given set of points also results in a dual graph made of triangles connecting those points: this is the *Delaunay triangulation* [83]. Delaunay triangles have some specific properties that single them out among other possible triangulations. In particular, they maximize the minimum angle in all triangles, thus tending to favor "fat" triangles and to avoid "thin," elongated ones.

Such Delaunay triangles, therefore, offer a complementary way of paving the time-frequency plane. An example of the Delaunay triangulation constructed on spectrogram zeros is given in Figure 15.10. This figure shows a distribution of triangles that, while not equilateral, are relatively similar, with few "thin" triangles. Indeed, with a distance between time-frequency points that is homogeneous all over the plane and neither too small nor too large, the isoperimetric inequality for triangles guarantees that the "fattest"

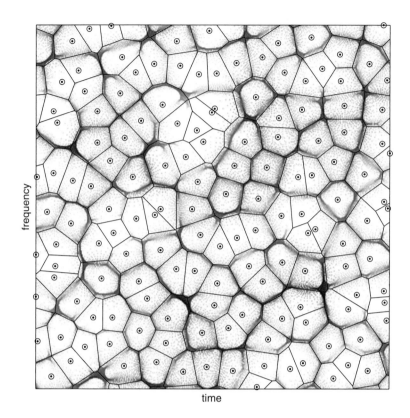

frequency

time

Figure 15.7 Reassigned spectrogram and Voronoi tessellation based on zeros. This figure replicates the example of Figure 15.4, with zeros highlighted as circles and the attached Voronoi tessellation superimposed as thin lines. It turns out that most of the reassigned energy localizes on the edges of the Voronoi cells.

Delaunay triangles should be nearly equilateral. If we take this argument further and assume for them an idealized mean model made of perfect hexagons (each of area 2π) that would pave the plane, the zeros would be located on a regular triangular lattice. A simple geometrical evaluation shows that, in such a case, the lattice would have a spacing of $(2/\sqrt{3})^{1/2} 2\pi \approx 1.075 \times 2\pi$, compared with the value $(1.138 \pm 0.316) \times 2\pi$ that resulted from the simulation that was mentioned previously.

Basic geometry shows us that a honeycomb structure of hexagons of area 2π is naturally attached to a regular lattice made of equilateral triangles of area π (homothetically to the local maxima case of Figure 15.9). This is supported by the simulation in Figure 15.11, where the estimated mean area is found to be $0.996 \times \pi$.

If the spatial point process of zeros was Poisson, the area distribution of Delaunay triangles would be Gamma distributed. Figure 15.11 shows that a Gamma fit to the empirical area distribution is only approximate. In particular, we can observe that the asymptotic decay is faster than the "Gaussian" decay ($\sim \exp\{-\beta a^2\}, \beta > 0$) that would occur with a Gamma distribution. This observation can be connected to an important (and defining) property of the Delaunay triangulation, namely that:

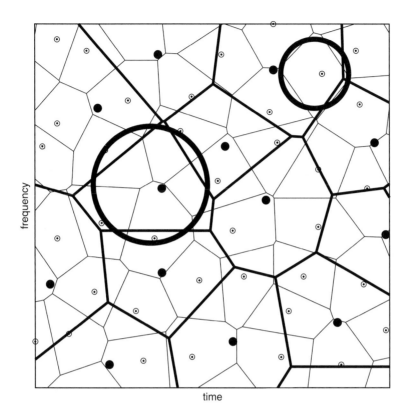

frequency

time

Figure 15.8 Voronoi tessellations based on spectrogram local maxima and zeros. The Voronoi tessellations attached to local maxima (black dots and thick lines) and to zeros (gray dots and thin lines) define two intertwined networks made of cells that, in the mean, are hexagonal and of respective areas $3 \times 2\pi$ to 2π. Circles with such areas are superimposed to the networks for the sake of visual comparison.

No point is inside the circumcircle of any Delaunay triangle.

This provides evidence of "voids" (or "holes") in the point distribution, as illustrated in Figure 15.12. It turns out that some statistics of those geometrical features can be obtained thanks to the theory of Gaussian analytic functions. In particular, it can be shown (see, e.g. [138]) that the probability $\mathbb{P}(N(r) = 0)$ that a disk of radius r contains no zero decays asymptotically as

$$\mathbb{P}(N(r) = 0) \sim \exp\{-\alpha r^4\}; \alpha > 0, \qquad (15.21)$$

thus suggesting that the probability of having Delaunay triangles with a large area should decay asymptotically as the *square* of the area. This interpretation is corroborated by the simulation results reported in Figure 15.11, where such a (log-)probability happens to be linear when plotted as a function of the squared area.

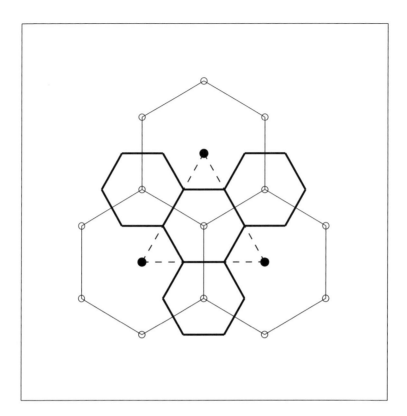

Figure 15.9 Honeycomb mean model in the white Gaussian noise case. In the mean model where spectrogram local maxima (dots) are located on a regular triangular lattice (dashed lines), the associated Voronoi cells (thin lines) are hexagonal and arranged in a honeycomb structure. Assuming that zeros (circles) are essentially located on the vertices of such hexagons, their Voronoi cells define themselves a secondary network of hexagonal cells (thick lines). By construction, the area of thin hexagons is 3 times that of thick ones, and twice the area of the lattice triangles.

This same property can be used further to characterize the *longest edge* of a Delaunay triangle in the complex white Gaussian noise case, by considering that the length of such an edge varies essentially as the diameter of the largest disk containing no zero. Assuming a decay as in (15.21) for the maximum edge length ℓ_M yields a probability distribution $p(\ell_M)$, whose right tail should be such that

$$\log p(\ell_M) \sim \beta - \alpha \left(\ell_M^4 + \frac{3}{\alpha} \log \ell_M \right); \beta \in \mathbb{R}. \qquad (15.22)$$

Figure 15.13 shows that this prediction is in good agreement with numerical simulations. While the core of the distribution is roughly Gaussian, its right tail obeys a faster decay that is well described by (15.22) with $\alpha = 1.5$.

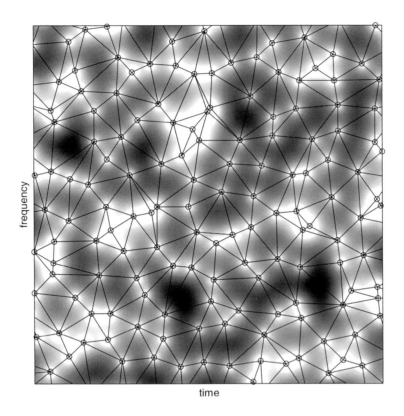

frequency

time

Figure 15.10 Spectrogram and Delaunay triangulation. This figure displays, on the same example of a spectrogram of complex white Gaussian noise as in Figure 15.4, the zeros highlighted as dots and the attached Delaunay triangulation superimposed as thin lines.

15.6   Signal Extraction from "Silent" Points

The study of spectrogram characteristics in the case of complex white Gaussian noise is an interesting field of pure research. It is also the prerequisite for dealing with "signal + noise" mixtures, with objectives such as signal detection and disentanglement or denoising. In such situations, the noise-only case becomes the null hypothesis H_0 in a binary model of the form

$$\begin{cases} H_0 & : y(t) = n(t) \qquad ; \\ H_1 & : y(t) = n(t) + x(t) \quad , \end{cases} \qquad (15.23)$$

where $y(t)$ is the observation and H_1 stands for the hypothesis in which some signal of interest $x(t)$ is superimposed to noise.

Deciding that the observation belongs to H_1 can therefore be achieved by contrasting the actual observation with what would be expected in a noise-only situation, e.g., by assessing some "outlier" behavior as compared to the nominal situation, where only noise would be present.

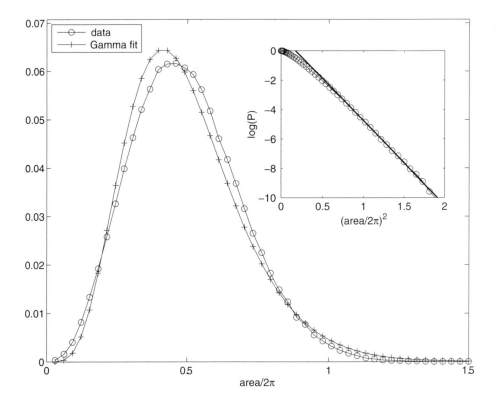

Figure 15.11 Delaunay triangles area distribution and "hole" decay probability. This figure displays the estimated probability density function of the area of Delaunay triangles constructed on zeros (circles), as well as its best Gamma fit (crosses). The discrepancy between data and fit is shown more precisely in the inserted diagram, which plots the (log-)probability of exceeding some given area as a function of the square of this area, with its linear fit superimposed.

> Since spectrogram zeros appear to carry most of the information of the analyzed observation, we can build such a signal extraction scheme by exploiting the knowledge we have gained about the distribution of zeros in the white Gaussian noise case.

The rationale of the approach is based on ideas introduced in [146]: in brief, the presence of a coherent component in some background noise disturbs the distribution of the spectrogram zeros. To understand the way this distribution is modified when some signal component is added to noise, we can use the following analogy. Focusing on the complex white Gaussian noise case, let us think of the time-frequency plane as a flat sheet and of the spectrogram zeros as marbles that are randomly, but homogeneously, distributed over this sheet. Within this picture, the addition of a signal component can be seen as a force that pushes the sheet from below. This force acts in places where the signal energy is located (e.g., along the time-frequency trajectory of the instantaneous frequency in the case of a chirp). Therefore, the surface of the sheet is no longer flat, thus

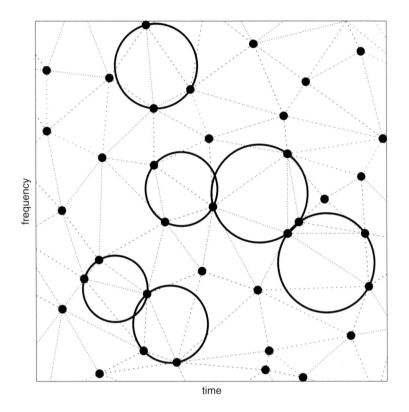

Figure 15.12 Delaunay triangulation and "holes." This figure displays spectrogram zeros (dots) in the case of a realization of complex white Gaussian noise, together with the associated Delaunay triangulation (dotted lines) and some circumcircles (thick lines) of arbitrarily chosen triangles. This evidences that the latter delineate "holes" in the point distribution.

repelling the marbles and creating a void whose border delineates the time-frequency domain of the signal. Because of the reproducing kernel of the analysis (and of the associated correlation structure), the repelled zeros cannot be arbitrarily close along this border but instead tend to organize to be as close as possible in an equidistributed way.

The key observation (illustrated in Figure 15.14), therefore, is that the Delaunay triangles constructed on the spectrogram zeros delineating a signal domain become more elongated than in the noise-only case, from which it follows that:

> The edge length is a distinctive feature that can be used to identify signal domains in the time-frequency plane.

When comparing the "constellations" of zeros, with and without the chirp superimposed, it appears that the modifications are *local*, in the sense that zeros are unaffected in noise-only regions. This can be seen as a natural consequence of the locality induced by the reproducing kernel of the analysis, and the key for making use of zeros in order to identify signal regions.

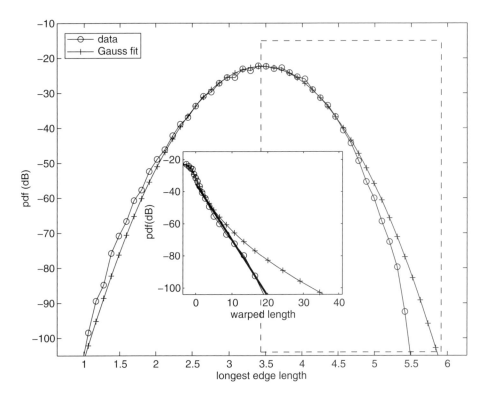

Figure 15.13 Longest edge length of Delaunay triangles. This figure displays the estimated probability distribution function (circles) of the longest edge of Delaunay triangles constructed on spectrogram zeros in the case of complex white Gaussian noise, together with its Gaussian fit (crosses). The discrepancy between data and fit is shown more precisely in the inserted diagram, which plots the log-probability density function within the dotted domain as a function of the warped length $\ell_M^4 + 2\log\ell_M$, with its linear fit superimposed.

Remark. The binary model (15.23) is quite general. From a practical point of view, it might seem restrictive to only consider the complex white Gaussian noise idealization for $n(t)$. What does happen when the background noise, although Gaussian, is colored (i.e., has a spectrum which is no longer flat)? It is difficult to give a general answer, due to the infinitely many ways of departing from a flat spectrum. Some insight can, however, be gained in the case of broadband noises whose spectrum has a smooth and "slow" variation. Indeed, a colored noise can be simply obtained by the convolutive action upon white noise of some filter with impulse response $h(t)$ or, in other words, by reshaping the flat spectrum by the multiplicative action of the transfer function $H(\omega)$ in the frequency domain. Assuming that the variation of $H(\omega)$ is sufficiently slow on the essential support of $G(\omega)$ (the Fourier transform of the Gaussian analysis window $g(t)$), we can make the local approximation $H(\xi)G^*(\xi - \omega) \approx H(\omega)G^*(\xi - \omega)$, so that

$$S_{n_h}(t, \omega) \approx |H(\omega)|^2 S_n(t, \omega) \tag{15.24}$$

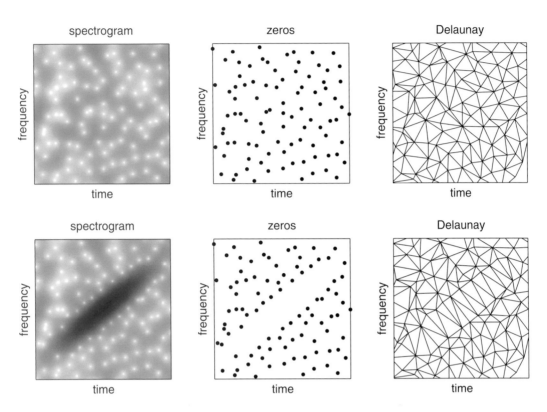

Figure 15.14 H_0 vs. H_1. This figure displays the spectrograms (left column), the corresponding zeros (middle column), and the associated Delaunay triangulations (right column) in the noise-only case H_0 (top row) and in the H_1 case (bottom row) where a linear chirp is added to the same noise realization, with a signal-to-noise ratio of 20 dB. © 2016 IEEE. Adapted, with permission, from [156].

if $n(t)$ stands for the input noise and $n_h(t)$ for its filtered version by $h(t)$. Under such an assumption, it follows that the zeros of the spectrogram remain almost unchanged after filtering, provided that the transfer function $H(\omega)$ introduces no further zeros. In a discrete-time setting, this is typically the case for autoregressive (AR) filters with poles not too close from the unit circle. An example is given in Figure 15.15, with an AR(2) filter whose (complex-conjugated) poles have a magnitude 0.8. The same behavior would be observed as well for other classical and versatile models of broadband noises such as *fractional Gaussian noise*, whose log-spectrum is a *tilted* version of the flat spectrum of white Gaussian noise. Of course, the approximation (15.24) would fail if the spectrum happened to be made of sharp, narrow peaks, making such a noise contribution indistinguishable from a tonelike signal component.

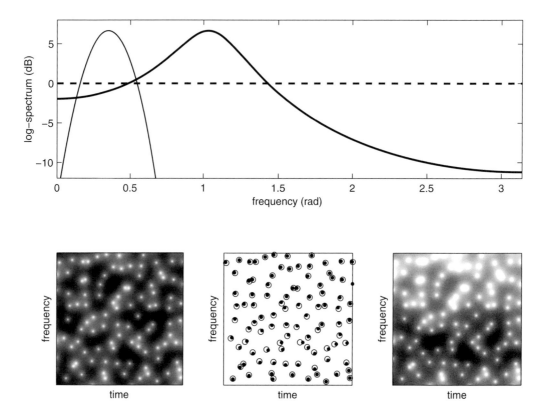

Figure 15.15 Spectrogram zeros in the case of a complex colored Gaussian noise. The top row of this figure displays an example of theoretical log-spectrum for a broadband noise (thick line) generated by the filtering of complex white Gaussian noise (dotted line) with an AR(2) filter, together with the log-spectrum of the Gaussian analyzing window (thin line). The bottom row presents the spectrograms of one realization of complex white Gaussian noise (left) and of its AR(2) filtered version (right), together with the superposition of the corresponding zeros (circles and dots, respectively) in the middle diagram.

All the elements discussed so far pave the way to a zero-based methodology aimed at identifying and extracting signal components embedded in broadband noise. This can be summarized by the following steps:

1.   Perform a Delaunay triangulation over the spectrogram zeros;
2.   Identify outlier edges with respect to the length distribution in the noise-only case;
3.   Keep triangles with at least one outlier edge;
4.   Group such triangles adjacently in connected, disjoint and labeled domains D_j;
5.   Multiply the STFT with labeled 1/0 masks attached to the different domains;
6.   Reconstruct the disentangled components $x_j(t)$, domain by domain, by using STFT reconstruction formulæ.

As for point 2, a threshold has to be chosen for assessing the "outlier" character of the longest edge of a Delaunay triangle. This can be done by using the results discussed around Figure 15.13 about the asymptotic decay of the probability density function. It is up to the user to determine the effective value of the threshold for guaranteeing a desired false alarm probability: as an example, the reference value 5 is such that $\mathbb{P}\{\ell_M \geq 5\} \lessapprox 10^{-3}$.

Concerning then point 6, one can use either the standard reconstruction formula:

$$x(t) = \iint_{D_j} F_x(\tau, \xi)(\mathbf{T}_{\tau\xi}g)(t)\, d\tau \frac{d\xi}{2\pi}. \tag{15.25}$$

or the simplified one, which had already been used for synchrosqueezing (see (10.15)):

$$x(t) = \frac{1}{g(0)} \int_{d_j(t)} F_x(t, \omega) \frac{d\omega}{2\pi}, \tag{15.26}$$

where $d_j(t)$ stands for the frequency section of D_j at time t.

The overall procedure is illustrated in Figures 15.16 and 15.17, on the simple example of a Hermite function embedded in a realization of the colored Gaussian noise considered in Figure 15.15.

Among the potential interests of the zeros-based filtering, one can mention that identifying connected domains on the basis of outlier edges in the Delaunay triangulation permits a simultaneous *labeling* of those domains. This contrasts with more classical approaches, which would be based on thresholding the magnitude of the transform, resulting in possibly disjoint domains that would require some post-processing so as to disentangle them. Magnitude thresholding also results in some necessary shrinkage of the identified domains, while anchoring their borders to zeros permits the signal energy embedded in noise to be captured in a more intrinsic way. Of course, the identification based on zeros relies on a threshold too (edges in the plane in place of magnitudes) and it is subject to the usual trade-off between detection and false alarm.

The zeros-based approach to time-frequency filtering can be seen as a kind of *a contrario* method: the existence of "something" is inferred from the knowledge of where there is "nothing."

> In audio terms, *sound is what exists between silences*

(as peace was defined by the French writer Jean Giraudoux as the time interval between two wars...).

Remark. This dual perspective between existence and nonexistence can be found in other contexts. In time series analysis, for instance, it is well known that an autoregressive model of order 1 can be equivalently represented as a moving-average model of infinite order, and vice versa. Recalling that the former model is all-pole and the latter model is all-zero, having one pole contribution can be seen as having an infinity of zero contributions, except at the very place where the pole is supposed to be located. In this sense, one pole is the absence of a zero. Conversely, one contribution can be seen as

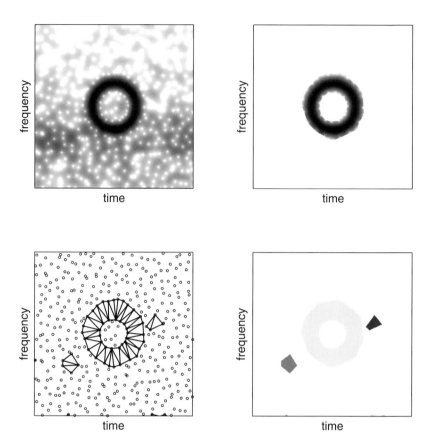

Figure 15.16 Zeros-based filtering. This figure illustrates the way spectrogram zeros can be used for filtering out signals from noise. Top left: spectrogram of a modulated Hermite signal of order 25 embedded in complex colored Gaussian noise, with a signal-to-noise ratio of 10 dB. Bottom left: constellation of zeros with outlier edges of the Delaunay triangulation highlighted with solid lines. Bottom right: Time-frequency domains identified from the Delaunay triangulation (three domains, with different gray colors, in the present case). Top right: Masked spectrogram after multiplication by a 1/0 function whose support is that of the light gray domain of the bottom right diagram.

resulting from pole contributions everywhere, except at the zero location. In both cases, a defining feature is equivalent to the absence of its opposite.

In the zeros-based approach, time-frequency domains attached to a signal component are directly anchored to spectrogram zeros. As shown in Figure 12.5, they provide an effective way of estimating the area "clear of zeros" corresponding to basins of attraction. Those basins can, however, be considered as well for signal disentanglement and reconstruction. In the case of the noisy Hermite signal, the contour-based identification of such basins leads to the segmentation displayed in Figure 15.18. The annular domain of the Hermite signal is globally identified, yet split into two main subdomains, thus

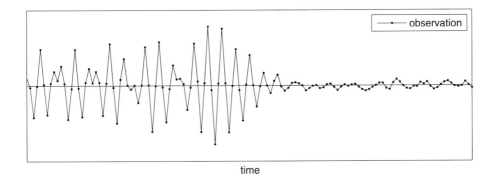

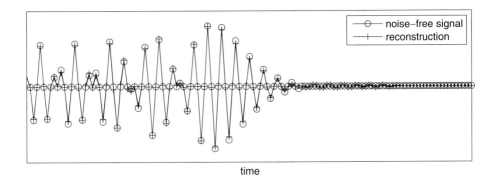

Figure 15.17 Zeros-based reconstruction. This figure illustrates an example of reconstruction after the zeros-based filtering of signals described in Figure 15.16. Top diagram: detail of the noisy observation, with an input signal-to-noise ratio of 10 dB. Bottom diagram: corresponding values of the noise-free signal and the reconstruction, with an output signal-to-noise ratio of 22.3 dB.

requiring some supervised post-processing. Masking the STFT by the union of these two domains would result in a reconstruction performance not far from that obtained with the zeros-based approach. Interested readers are referred to, e.g., [117] for further quantitative comparisons.

15.7 Universality

To conclude this chapter about zeros, let us discuss how they are in fact characteristic points of a spectrogram with some universal and well-defined structure in their vicinity. Indeed, with the notations used in (12.4) and (12.6), we can rewrite the Bargmann transform $\mathcal{F}_x(z)$ as a function of its magnitude and phase, i.e., as:

$$\mathcal{F}_x(t, \omega) = \mathcal{M}_x(t, \omega) \exp\{i\Phi_x(t, \omega)\}, \tag{15.27}$$

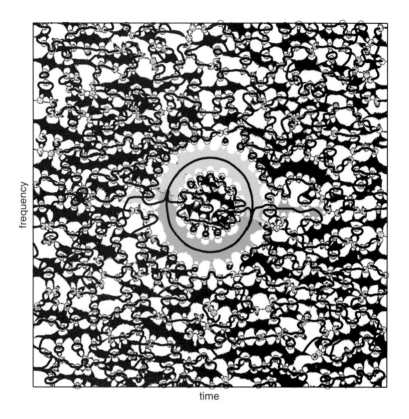

frequency

time

Figure 15.18 Contour-based basins. This figure illustrates the contours (full lines) and the contour-based identification of basins (here identified with different shades of gray), in the example considered in Figure 15.16. The annular domain of the Hermite function is globally identified, yet split into two sub-domains.

where $\Phi_x(t, \omega)$ stands equivalently for the phase of either the Bargmann transform $\mathcal{F}_x$ or the STFT F_x. It has been shown (see (12.7)–(12.8)) that this phase and the log-magnitude $\log \mathcal{M}_x(t, \omega)$ form a Cauchy pair:

$$\frac{\partial \Phi_x}{\partial t}(t, \omega) = \frac{\partial}{\partial \omega} \log \mathcal{M}_x(t, \omega); \tag{15.28}$$

$$\frac{\partial \Phi_x}{\partial \omega}(t, \omega) = -\frac{\partial}{\partial t} \log \mathcal{M}_x(t, \omega), \tag{15.29}$$

with the result that phase properties are entirely determined by the knowledge of the only magnitude. This phase gradient $\nabla \Phi_x = \mathrm{Im}(\nabla \log \mathcal{F}_x)$, either in its direct form or in its equivalent expression via the log-magnitude, is the key ingredient of reassignment. In this respect,

> The acknowledged fact that spectrogram zeros are *repellers* for the reassignment vector field happens to be intimately related to the companion property of zeros as *singularities* of the phase gradient.

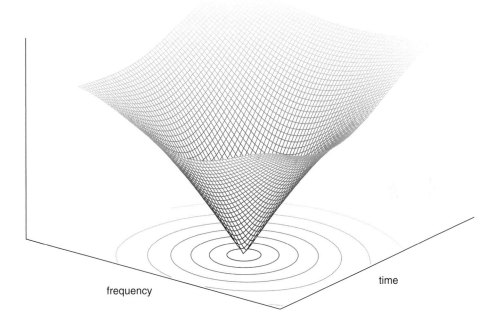

Figure 15.19 STFT magnitude in the vicinity of a zero . This figure illustrates the linear behavior of the STFT magnitude in the vicinity of a zero as predicted by (15.30).

More precisely, thanks to the Weierstrass-Hadamard factorization (15.3), the magnitude of the Bargmann transform (and, hence, of the STFT) can be approximated in the vicinity of a given zero z_n by

$$\mathcal{M}_x(t, \omega)|_{t \sim t_n, \omega \sim \omega_n} \propto |z - z_n| = \sqrt{(\omega - \omega_n)^2 + (t - t_n)^2}. \tag{15.30}$$

This is illustrated in Figure 15.19.

Combined with (15.28)–(15.29), this leads directly to the singular behavior of the phase partial derivatives, in the form of a *hyperbolic divergence*:

$$\left.\frac{\partial \Phi_x}{\partial t}(t_n, \omega)\right|_{\omega \sim \omega_n} \sim \frac{1}{\omega - \omega_n}; \tag{15.31}$$

$$\left.\frac{\partial \Phi_x}{\partial \omega}(t, \omega_n)\right|_{t \sim t_n} \sim \frac{1}{t_n - t}. \tag{15.32}$$

An illustration is given in Figure 15.20.

This striking behavior, which corresponds to the systematic twin appearance of two sharp peaks of opposite sign, has been given a simple justification here based on the generic factorization of the Bargmann transform [110], thus justifying its universality in the circular Gaussian case. Its analysis can also be given a complete analytic treatment

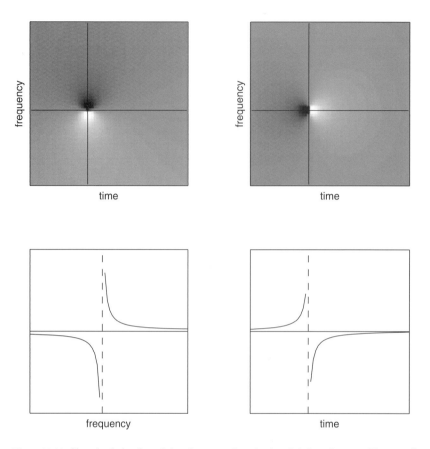

Figure 15.20 Singular behavior of the phase gradient in the vicinity of a zero. The top diagrams display the phase gradient (left: in time, and right: in frequency) in the vicinity of a STFT zero, which is located at the intersection of the two superimposed full lines. The bottom diagrams plot the corresponding sections of the gradient surfaces, illustrating the hyperbolic divergences predicted by (15.31) and (15.32).

[147] that carries over to more general – even noncircular – windows, provided they are smooth enough.

15.8   Singularities and Phase Dislocations

It readily follows from the complex-valued nature of the STFT that the phase of the transform reads

$$\Phi_x(t,\omega) = \tan^{-1}\left(\frac{\text{Im}\{F_x(t,\omega)\}}{\text{Re}\{F_x(t,\omega)\}}\right), \tag{15.33}$$

hence an indeterminacy when both the real and imaginary parts simultaneously vanish, i.e., at zeros. When looking in parallel at the magnitude and the phase of the STFT, as shown in Figure 15.21, it clearly appears that the phase takes on a specific, branching

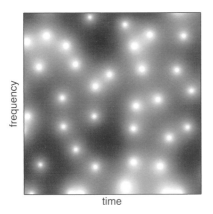 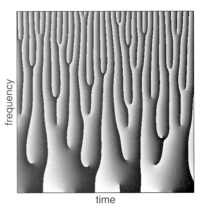

Figure 15.21 Spectrograms zeros and phase singularities. This figure illustrates that, when the magnitude of the STFT (left diagram) vanishes, the phase (right diagram) experiences an indeterminacy, which results in a branching structure.

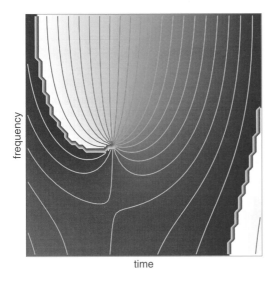

Figure 15.22 STFT phase in the vicinity of a zero. This figure zooms in on the most bottom left zero in Figure 15.21 and superimposes isocontours to the gray-coded phase surface.

structure at those precise locations. Figure 15.22 zooms in on the phase at a particular zero, thus showing the singularity structure in a finer way. From a very general point of view, the singularity that is observed for the derivatives of the STFT phase around zeros is related to a "phase dislocation" mechanism that is reminiscent of what may happen in wave trains, as discussed in [148].

One of the possible ways the magnitude goes to zero is illustrated in Figure 15.23, where a sequence of successive time slices of the (real and imaginary parts of the) STFT – in the neighborhood of a frequency where a zero magnitude occurs in the considered timespan – is complemented by the associated phase diagram (sometimes

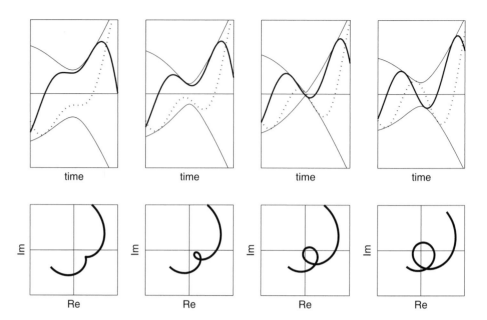

Figure 15.23 STFT zeros and phase singularities. The top diagrams plot four slices of the STFT as a function of time, for four neighboring frequencies (in increased order, from left to right). The real part, the imaginary part, and the envelope are plotted as thick solid lines, dotted lines, and thin solid lines, respectively. The bottom diagrams are the corresponding Argand (or Fresnel) diagrams, in which the imaginary part is plotted as a function of the real part. The figure illustrates that the vanishing of the envelope in the time domain (third column from the left) corresponds to the "touching" of the origin by the curve in the phase diagram. Further increasing the analyzed frequency results in the creation of an extra loop in the phase diagram, i.e., one more oscillation in the time domain.

attached to the names of Argand or Fresnel) in which the imaginary part is plotted as a function of the real part.

In such a phase diagram, the quasi-monochromatic oscillation of some STFT time slice is described by a closed loop that circles the origin of the plane in a quasi-periodic fashion, the mean period being directly related to the inverse of the considered frequency. These two representations of the same phenomenon (time slices and phase diagram) clearly show that the vanishing of the magnitude corresponds to the very time instant where the curve that describes the local oscillation of the waveform "touches" the origin of the complex plane. When the reference frequency is progressively increased, one new inner loop is created in the trajectory, and this loop eventually leads to one more oscillation in the waveform when it happens to cross the origin.

The situation is therefore pretty much similar to what is discussed in [148], although there are some differences. Indeed, the phase dislocations described in [148] are the result of an interference process involving distinct, quasi-monochromatic wave trains – such as the ones created by reflections (echoes) of one emitted waveform on some rough surface – which result in a superposition of delayed replicae coming from slightly different directions in space. However, the interfering waveforms have the same frequency content, and as a result, this is also the case for their superposition; the phase dislocation corresponds only to a dramatic change in the locations of the extrema, but does not affect how far apart they are.

The situation is somewhat different in the STFT case, where dislocations cannot be viewed as the result of some interference between those wave trains whose time lag would result from different propagations in space. An interpretation that better suits the STFT situation can be proposed thanks to the reproducing formula (6.3).

According to this relation, any time slice of the STFT at some given frequency can be viewed as resulting from the interfering superposition of similar time slices attached to neighboring and slightly different frequencies. Given a fixed timespan, the creation of one extra oscillation is therefore intimately linked, within this picture, to an increase in frequency, each passing of the phase diagram trajectory through the origin corresponding to a phase dislocation with a branching structure. As expected, exploration of higher and higher frequencies goes together with more and more closely spaced isophase contours in the time-frequency plane: this is made possible thanks to jumps at those dislocation points where the magnitude of the transform vanishes.

16 Back to Examples

Some examples of actual data were introduced in Chapter 2 to serve as a motivation for the development of time-frequency methods aimed at their analysis. The subsequent chapters have been taken up with introducing, discussing, and interpreting these methods, and we have now reached the point where we can return to data. This is the purpose of this chapter, which will provide examples of the use of dedicated time-frequency techniques, and the role they can play in both analysis and processing of "small" signals in highly nonstationary contexts.

16.1 Gravitational Waves

In Chapter 2, we said that detection on Earth of gravitational waves can only be expected from extremely energetic astrophysical events. In this respect, a prime candidate is the merger of binary systems formed by neutron stars or black holes rotating around each other.

Gravitational waves are sometimes referred to as "ripples in spacetime." Those tiny vibrations induce a differential effect on the arm length of the interferometers aimed at their detection. As received by the interferometric detectors, a gravitational wave gives rise to a displacement of the fringe pattern that follows the oscillations of the impinging wave. In the case of coalescing binaries, dissipation, rotation, and conservation of angular momentum imply that the objects become closer and closer, thus speeding up rotation, up to merging. The overall result is that gravitational waves *signals* – which are precisely measured via the oscillations of the fringe pattern – are expected to behave as *chirps*, with increasing instantaneous frequency and amplitude.

Model: This phenomenological and qualitative description can be given much more precise forms. In a first (Newtonian) approximation, an explicit expression has been established for the typical waveform expected to result from the coalescence of a binary system (see, e.g., [149], [150], or [151]):

$$x(t; t_0, d) = A\,(t_0 - t)^{-1/4}\,\cos\left(2\pi\,d\,(t_0 - t)^{5/8} + \varphi\right)\mathbf{1}_{(-\infty, t_0[}(t). \qquad (16.1)$$

In this expression, t_0 is the coalescence time, φ some phase reference, and d and A are constants that mainly depend on the individual masses of the objects as well as of other geometrical quantities such as the Earth-binary distance or the relative orientation between the wavefronts and the detector.

Remark. Of course, the model (16.1) is an idealization, and it has to be understood as an approximation for the *inspiral* part of the coalescence, with a validity that can be questioned when t becomes too close from t_0. The divergence which occurs at time $t = t_0$ is unphysical and has to be regularized in some way. This can be achieved by more elaborated analyses such as the "Effective One-Body" approach [152] which not only permits us to prevent divergence, but also models the part of the signal where the two bodies merge into a single black hole, and the ringdown of this black hole as it relaxes to equilibrium.

To be more precise, if we consider that the binary is made of two objects of individual masses m_1 and m_2, one can introduce the "total mass" $M = m_1 + m_2$ and the "reduced mass" μ such that $\mu^{-1} = m_1^{-1} + m_2^{-1}$. Using these two quantities, one can then define the "chirp mass" $\mathcal{M} = \mu^{3/5} M^{2/5}$. Following [150] and [153], we have

$$d \approx 241 \, \mathcal{M}_\odot^{-5/8}, \tag{16.2}$$

with $\mathcal{M}_\odot = \mathcal{M}/M_\odot$, in which $M_\odot$ stands for the solar mass. For an optimal relative orientation between the detector and the binary, we have furthermore

$$A \approx 3.37 \times 10^{-21} \, \frac{\mathcal{M}_\odot^{5/4}}{R}, \tag{16.3}$$

where R is the Earth-binary distance, expressed in Mpc.

It follows from (16.1) that, for instantaneous frequency, the gravitational wave chirp has the power-law form:

$$f_x(t) = \frac{5d}{8} \, (t_0 - t)^{-3/8} \, \mathbf{1}_{(-\infty, t_0[}(t) \tag{16.4}$$

which, in theory, glides over the whole half-line of positive frequencies, albeit in a highly nonlinear way. In practice, the observable range of frequencies is limited, offering a "window" which extends roughly between some tens and some thousands of Hz, lower frequencies being dominated by seismic noise and higher ones by photon noise.

Thanks to (16.4), it is possible to convert the frequency window into a time window and, in order to have clearer ideas about observability, we can give some orders of magnitude. If we let T be the effective duration of the chirp within the observation window, i.e., the length of time between when the chirp "enters" the frequency window $f \leq f_m$ and coalescence, we find that:

$$T \approx 6.35 \times 10^5 \, f_m^{-8/3} \, \mathcal{M}_\odot^{-5/3}. \tag{16.5}$$

Letting $f_m \approx 30$ Hz, it turns out that $T \approx 250$ ms for the (astrophysically reasonable) assumption that $\mathcal{M}_\odot \approx 30$ (and $T \approx 500$ when $\mathcal{M}_\odot \approx 20$). In other words, while the coalescence process of a binary might have started in a fairly remote past, the gravitational wave chirp we can expect to detect is a *very transient* signal, with an observable lifetime of a fraction of a second!

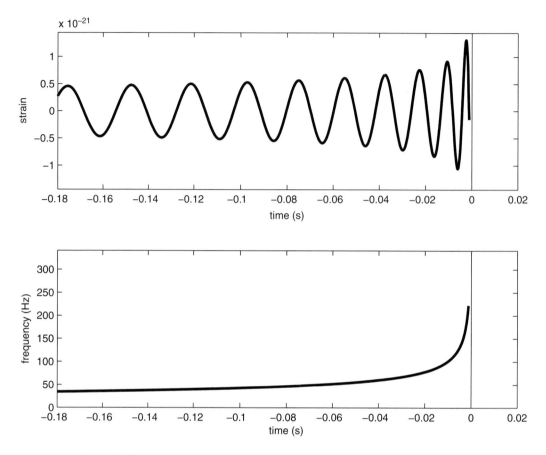

Figure 16.1 Gravitational wave model in the case of a coalescing binary system, according to (16.1). Top diagram: waveform. Bottom diagram: instantaneous frequency. In both diagrams, the coalescence time is arbitrarily fixed at $t = 0$. The model parameters have been chosen to roughly mimic the actual event GW150914, i.e., $M_\odot = 30$ and $R = 800$ Mpc.

Those findings corroborate the values reported in Figures 2.1 and 2.2. For the sake of comparison, an example of the model (16.1)–(16.4) is displayed in Figure 16.1, where the parameters and the time-frequency characteristics have been chosen so as to roughly mimic the actual observation of the event GW150914.

Remark. The amplitude of the waveform corresponds to the "strain" or, in other words, to the dilation/shrinkage of distances by unit length: it is remarkable to see how tiny this relative displacement is, namely of the order of 10^{-21}! This gives a measure of the formidable technological challenge to be faced for detecting a gravitational wave.

Time-frequency detection: Gravitational wave signals carry some physical information about the system from which they originate, but before processing the received waveforms in order to *estimate* the physical parameters they encode, the very first challenge is to *detect* them. Different strategies can be developed, depending in particular on the amount of a priori assumptions that can be made about the waveforms to detect and the background noise in which they are embedded. When formulating the detection problem as a binary hypothesis test such as in (15.23), two main options are offered:

1. assume that a reasonable model is available for the waveform $x(t)$ to be detected from the observation $y(t)$, with white Gaussian noise $n(t)$ as background noise: this typically leads to a *matched filtering* approach aimed at accepting H_1;
2. make no specific assumption about $x(t)$, except that its presence transforms the observation $y(t)$ into an outlier as compared to the noise-only case: this leads to an *a contrario* approach aimed at rejecting H_0.

As for the effective detection of GW150914, both strategies have been followed by the LIGO-Virgo collaboration [5]. Option 1 was based on a bank of matched filters which, in a first approximation, were supposed to cover the parameter space in models such as (16.1). Option 2 made no such assumption, but took advantage of the existence of two measurement devices operating simultaneously (the two interferometers located in Hanford, WA and Livingston, LA), with the idea of using coincidences for ascertaining decision.

To be more precise, this second approach (developed by Serguëi Klimenko and his group, and referred to as "Coherent Wave Burst" [134]) was basically a time-frequency one, with the following rationale. Whatever its fine structure, chirp or not, a gravitational wave signal is expected to be a transient waveform creating some localized excess of energy in the time-frequency plane. Identifying such an excess that would occur in a coherent way for the two detectors – which are sited at locations approximately 3,000 km apart, and can therefore be considered as independent in terms of measurement – should therefore be a strong indication of an event. From a technical point of view, the signals received at the two detectors are independently projected onto a *Wilson basis* [154], which is a variant of discrete Gabor expansion using the so-called *Meyer wavelet* [20] as a short-time window. In such a decomposition, transient signals are sparse and characterized by significant values for a few coefficients only. Based on a statistical knowledge of the fluctuations of coefficients in the noise-only situation, outlier coefficients are first identified. A rejection test is then constructed by comparing those outlier coefficients in a coherent way (in both magnitude and phase).

Remark. In [5], it is reported that the first detection was achieved by Option 2 ("Coherent Wave Burst") and not by Option 1 ("matched filtering"), while a subsequent analysis proved that the underlying model was quite accurate. The reason is twofold. First, GW150914 was an event that was much more energetic than what had been envisioned in tentative scenarios, resulting in a high signal-to-noise ratio and making it "easier" for Wilson coefficients to emerge from noise. Second, this high energy was related to the fact that the event corresponded to the coalescence of two black holes that were much more massive than imagined (about 30 solar masses each) and whose parameters, as a result, were outside of the grid used for online matched filtering searches that were primarily targeting low-mass, neutron star binaries ... It is worth noting that a second event was detected some weeks later [155]. In this case, less massive objects were involved (about 20 solar masses each), the signal-to-noise was not as good, and matched filtering was successful, while Coherent Wave Burst was not.

Time-frequency filtering: After a fair amount of preprocessing [5], the observations are still noisy (see Figure 2.1 for the Hanford signal) and call for a cleaning that can be advantageously performed in the time-frequency plane due to the chirping nature of the waveforms. While this was achieved in [5] by using wavelet techniques, we can also imagine following this up by using the approach developed in Chapter 15 on the basis of spectrogram zeros and Delaunay triangulation [156].

In the cases of both Handford signals and Livingston signals, the observation (which is available from https://losc.ligo.org/events/GW150914/) is made of 3,441 data points, sampled at the rate of 16,384 Hz. Time-frequency analyses (STFTs, spectrograms and their reassigned versions) have been performed with a time step of 8 samples and a frequency zoom in on the lower part of the spectrum by a factor of 24. This results in a time-frequency matrix of size 431×85, covering the effective time-frequency domain [0.25 s, 0.46 s] $\times$ [0 Hz, 341 Hz]. The short-time window is a Gaussian whose size (1,023 samples) has been chosen to be almost "circular" for the retained sampling of the time-frequency plane. In order to fix the threshold for discriminating outlier edges in the Delaunay triangulation, the edge distribution has been estimated from a number of surrogate spectrograms computed on noise-only signals, under the same conditions as the ones used for the signal under test.

The procedure is illustrated in Figure 16.2, in the case of the Hanford signal. A similar analysis can be performed on the Livingston signal.

Comparisons: If both Hanford and Livingston signals do correspond to the same event, it should be possible to make them coincide after a proper shift related to the propagation, at the speed of light, of the gravitational wave from one detector to the next.

Due to the relative orientation of the detectors, this transformation involves a sign flip (i.e., a phase shift of π radians) as well as a time delay of at most 10 ms, given that the two detectors are spaced about 3,000 km apart and that gravitational waves travel at the speed of light, i.e., 3×10^8 m/s. This expectation is clearly supported by the top diagram of Figure 16.3, in which the cross-correlation between the two filtered waveforms is plotted. The shape of this cross-correlation function is nearly symmetric, with a negative peak off the origin. The fact that this peak is negative and that it almost attains the value of -1 proves the sign flip that is expected in between the waveforms due to the relative positioning of the detectors. Furthermore, the time delay at which the peak is located from the origin turns out to be 6.8 ms. (Note that the detection in Hanford follows that in Livingston, which suggests that the event took place in the southern hemisphere.) The bottom diagram of Figure 16.3 displays the similarity between the two filtered waveforms, once properly synchronized according to the previously-mentioned shifts as well as to an *ad hoc* amplitude adjustment.

Besides observed data, results of numerical relativity computations are also available [5], and thus amenable to comparisons between theory and measurements. Such a comparison is reported in Figure 16.4 for both Hanford and Livingston signals, and shows an amazing agreement in both cases.

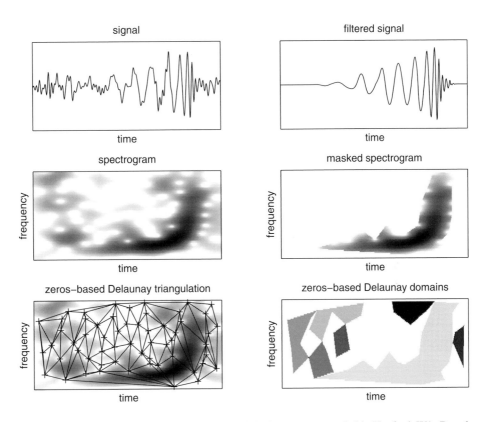

Figure 16.2 Time-frequency filtering of the GW150914 event, as recorded in Hanford, WA. Based on the STFT/spectrogram (middle left) of the noisy observation (top left), a Delaunay triangulation is computed on zeros (bottom left). This permits us to identify potential signal domains by concatenating triangles with edges longer than in the noise-only case (bottom right). Retaining one of those domains (namely, the light gray one in the present case) defines a 1/0 mask which, when applied to the STFT/spectrogram (middle right) allows for a denoised reconstruction by inverting the transform over this domain (top right). Axes for the waveforms are as in Figure 2.1 (and axes for the time-frequency diagrams are as in Figure 2.2). © 2016 IEEE. Adapted, with permission, from [156].

Time-frequency matched filtering: Once the observations have been filtered, one can go further and estimate the chirp parameters under the assumption that the inspiral part of the coalescence (before it attains its maximum and then evolves in a different ring-down mode) follows the approximation (16.1), which describes the way the amplitude and the instantaneous frequency diverge when approaching the coalescence time t_0.

This evolution is controlled by the "chirp rate" d, which happens to be related to the so-called "chirp mass" $\mathcal{M}_\odot$ according to (16.2). Given this model, two main parameters – namely t_0 and $\mathcal{M}_\odot$ – must therefore be estimated.

The classical approach for such a task is to make use of a bank of matched filters [150] but, as suggested in [153] and [157], an (approximate) equivalent formulation can be given in the time-frequency plane. The rationale is that, for a proper choice of time-frequency representation, the signature of a gravitational wave chirp such as

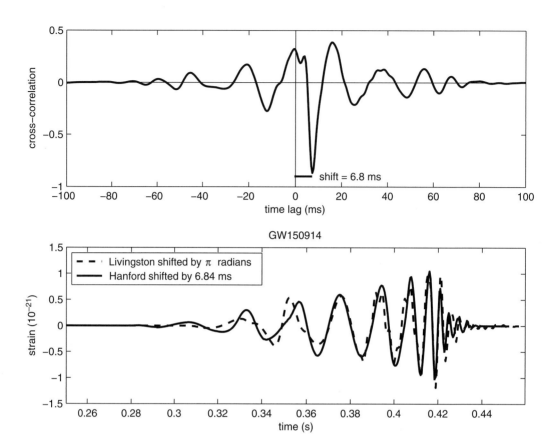

Figure 16.3 Comparison of GW150914 signals, as recorded in Hanford, WA and Livingston, LA. Top diagram: cross-correlation between the two filtered waveforms. Bottom diagram: superposition of the two waveforms, after suitable shifts in time and phase, and an *ad hoc* amplitude adjustment.

(16.1) can be made highly localized along the instantaneous frequency law given by (16.4). This is actually the case with reassigned spectrograms which, as detailed in [153], can be used to form an approximately optimal "time-frequency matched filter" by performing a (weighted) *path integration* along trajectories that precisely correspond to the instantaneous frequency model (16.4).

A joint estimation of t_0 and $\mathcal{M}_\odot$ can therefore be achieved as [153]:

$$(\hat{t}_0, \hat{\mathcal{M}}_\odot) = \arg \max_{(t_0, \mathcal{M}_\odot)} \int_{\mathcal{L}(t_0, \mathcal{M}_\odot)} \hat{S}_y(t, f) f^{-2/3}, \qquad (16.6)$$

with

$$\mathcal{L}(t_0, \mathcal{M}_\odot) = \left\{ (t, f) \mid t_0 - t = 6.35 \times 10^5 \, \mathcal{M}_\odot^{-5/3} \, f^{-8/3} \right\}. \qquad (16.7)$$

An example of this strategy is reported in Figure 16.5 in the case of the Hanford signal. The estimated value for the chirp mass was found to be about 28 solar masses, which is pretty consistent with what is reported in [5].

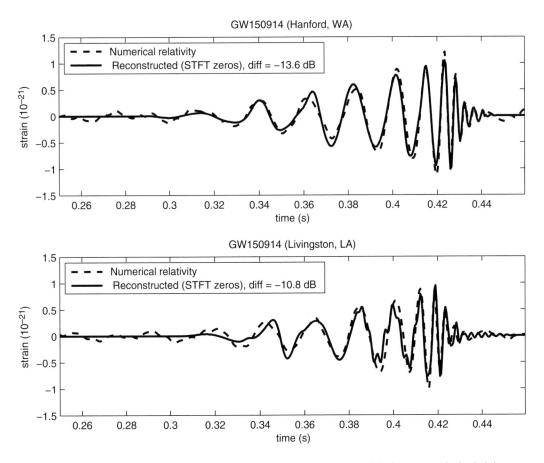

Figure 16.4 Comparison of filtered GW150914 signals with models from numerical relativity. © 2016 IEEE. Adapted, with permission, from [156].

16.2 Bats

In Chapter 2, we saw some examples of bat echolocation calls that revealed a chirping structure. We will now elaborate further on those preliminary findings, in a time-frequency perspective supported by some of the tools and interpretations developed in subsequent chapters.

Remark. Up to now, we have referred to "bats" somewhat loosely, but speaking about "bats" in general and in a comprehensive way would definitely be out of place in the specific context of this chapter. There are more than 1,200 species of bats; most of these are echolocating bats that use a variety of waveforms, which may not only differentiate between groups, but also between individuals within a given group. However, if we concentrate on the so-called *Vespertillonidae* ("evening bats"), which are the most common family of microchiropterae, we observe a number of common features such as the use of short, transient, descending chirps. This family encompasses members such as *Myotis lucifugus* (the "little brown bag" that was the subject of the first investigations by George W. Pierce and Donald W. Griffin [25]), *Myotis mystacinus* (the "whiskered

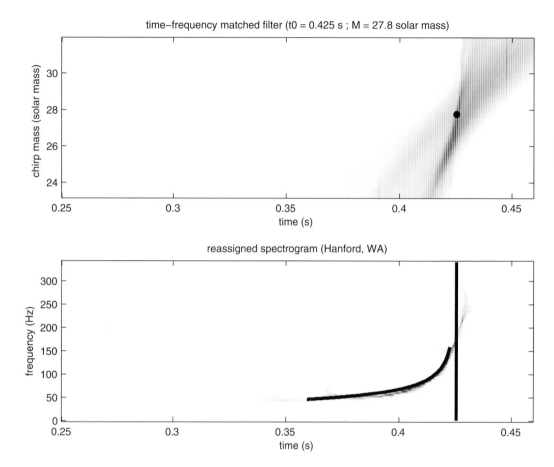

Figure 16.5 Time-frequency matched filtering of the filtered GW150914 Hanford signal. Top diagram: output of the "time-frequency matched filter" implemented by integrating the reassigned spectrogram of the observation along a set of the model time-frequency trajectories (16.4) indexed by possible coalescence times t_0 and chirp masses $M_\odot$ (the maximum value is indicated by a dot). Bottom diagram: reassigned spectrogram of the observation, with the best fit of the model, obtained with coordinates of the dot in the top diagram, superimposed as thick lines (the curve corresponds to the instantaneous frequency and the vertical line indicates the coalescence time).

bat"), *Esptesicus fuscus* (the "big brown bat"), and *Pipistrellus pipistrellus* (the "common pipistrelle"). The discussion in the following text will concern *Myotis mystacinus* specifically, but conclusions of a similar nature would apply more generally to the larger family of *Vespertillonidae*.

The material that can be used for a data analysis of bat echolocation calls and, more generally, of "animal sonar systems" [158] can be obtained in at least three ways. The first one is pure observation, with the recording of signals *in the field*, in an environment and under conditions that are "natural" to the animals. The second possibility is to confront bats with elementary tasks, but *in the lab* and under well-controlled protocols. The last option is offered by *neurophysiology*, with *in vivo* investigations of functionalities in relation with specialized areas in the bats' brains.

All three approaches have their pros and cons, and using all three provides access to complementary information that would not be available with only one of them. For instance, recordings in the field are of primary importance since they are nonintrusive and are supposed to provide access to signals that are actually used in natural conditions. The limitation, however, is that the operating situation can be hard to evaluate (is the bat routinely hunting? discovering a new environment? fighting against reverberation? managing to avoid other bats?...). Furthermore, it is very difficult to record signals from the target's perspective (i.e., what the bat actually emits) and echoes from the bat's perspective (i.e., what the bat actually receives back from the target); the result is that most recordings do not necessarily reflect what is actually emitted and/or received by the bat. Such limitations can be overcome in lab experiments, where bats are trained to "answer" questions related to single well-defined tasks, and where a much more controlled recording of the emitted waveforms is possible. Unfortunately, this also goes along with artificial situations that may depart from natural ones, casting doubt on the operational relevance of answers obtained in this manner. Finally, neurophysiological investigations can give some anatomical evidence about specialized areas and sensitivity to physical parameters controlling the emitted waveforms (such as, e.g., chirp rate), but they give no clue about their effective use in action.

In the field data: The typical pursuit sequence of *Myotis mystacinus* on which the analysis reported here is based was tape recorded in the field at Villars-les-Dombes (France), on May 6, 1982.[1]

The digitization was done with a sampling rate of 7.2 kHz after reducing the tape speed by a factor of 32 and a filtering within the band [250 Hz, 3.5 kHz]. This corresponds to an actual sampling rate of 230.4 kHz for the original analog data. After sampling, the total number of data points is 192,000 for the whole sequence, whose duration is therefore of 0.833 s.

Figure 16.6 displays this sequence in the time domain. It happens to contain about sixty separate signals, each of which is only a few milliseconds in length (this corresponds to a few hundred samples after digitization). Although it is difficult to gain much information about the individual signals from this representation, we can make a number of observations:

1. It is clear that the sequence undergoes some evolution concerning both the waveforms (which apparently differ between the beginning and the end) and the time rate of their emission, which is accelerated when approaching the end.
2. The amplitude is a decreasing function of time. This can be simply explained by considering that, in a successful hunting sequence, the predator-prey distance is progressively reduced until the prey is caught. In an echolocating scenario,

[1] This recording was part of the research program RCP 445 supported by CNRS (Centre national de la recherche scientifique, France) and developed at *ICPI Lyon* under the supervision of the late Bernard Escudié and Yvon Biraud, together with Yves Tupinier.

this in turn permits the volume level of the emitted signal to be reduced without sacrificing the signal-to-noise ratio that is necessary for detection.

3.   A related observation is that the first (most energetic) signals of the sequence are followed by secondary signals, which turn out to be caused by reverberation stemming from reflections off surrounding trees. Such environmental echoes should exist as well for the subsequent signals but, due to the low level of the emission, they are too faint to be observed. This is of course the same for actual echoes from the target (an insect).

4.   The typical hunting sequence consists of three main phases, namely *search* when looking for a prey item, *approach* when a contact has been made, and *attack* for the final part until the prey is caught and eaten. Approximate starts of those phases, highlighted by boxes, focus on sub-sequences to be studied further in greater detail.

A time-frequency reading: As announced and briefly explained in Chapter 2, a much more informative description of bat echolocation calls can be obtained by writing down

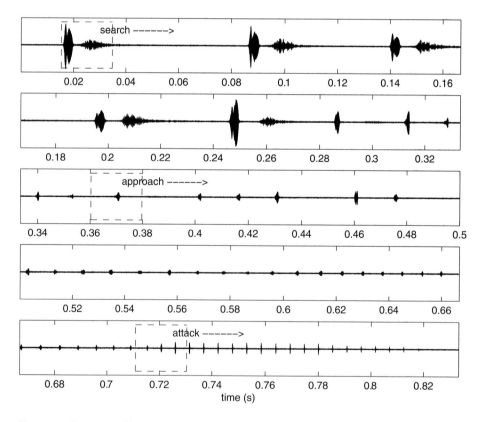

Figure 16.6 Sequence of bat echolocation calls in the time domain (to be read from left to right and from top to bottom). This whole hunting sequence of *Myotis mystacinus*, which lasts for less than 1 s, contains about 60 signals. The three boxes in dotted lines highlight the approximate start of the three phases that characterize a hunting sequence, namely "search," "approach," and "attack," in chronological order.

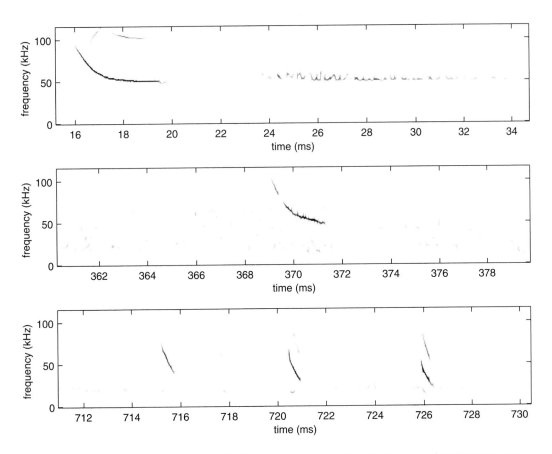

Figure 16.7 Bat echolocation calls in the time-frequency domain. Reassigned spectrograms are displayed for the three sub-sequences corresponding to the highlighted boxes in Figure 16.6, attached to the successive phases of "search," "approach," and "attack," from top to bottom. In all cases, the dynamic range is 30 dB.

their "musical score" with the aid of a time-frequency analysis. This is what is done in Figure 16.7, with a reassigned spectrogram chosen as a time-frequency energy distribution for the sake of high localization along the time-frequency trajectories of the chirps.

The successive sub-sequences corresponding to the three phases highlighted in Figure 16.6 display clear morphological changes in the chirping structure of the waveforms during the course of the hunting process. In all cases, the emitted waveform consists of a *descending chirp,* but the different phases call for the following phenomenological comments:

1.   "Search": With a duration of about 4 ms, the chirp sweeps over about one octave, from 100 to 50 kHz. The first half is characterized by a nonlinear instantaneous frequency law with upward concavity, which is followed in the second half by a frequency that almost stabilizes to a tone.

2.   "Approach": As compared to "search," the chirp is shorter in time, but it extends over roughly the same frequency band, with a similar nonlinear sweep. It behaves more or less as if the tone part had been erased.

3.   "Attack": Due to the acceleration in the emission rate, this sub-sequence includes three signals over a time span that is identical to those of phases 1 and 2. Signals are much shorter in time than before, with an almost linear instantaneous frequency. They also attain a lower frequency at the end of the chirp, with a clear transition in which a drastic drop in the bandwidth of the fundamental sweep is accompanied by the appearance of a (frequency doubling) harmonic component.

Why the observed chirps?: Describing bat chirps is one thing, but proposing explanations for why they take the forms they do would be more interesting. The ubiquity of the observed signals (and of the way they change within a hunting sequence) suggests that their design has been optimized to best deal with the situations that bats have to face for their survival.

Although it is biological, the echolocating system used by bats has many of the same components as manmade sonar systems; it therefore makes sense to try to model, if not explain, reported observations and findings in the manner used by signal and communication theory.

Remark. This argument can in fact be used in two opposite ways. On the one hand, we can imagine stating the "bat problem" as an engineering problem of signal design, and apply to it the machinery of models and solutions that have proved effective in manmade sonar systems. On the other hand, one can give way to the "bionic temptation," and think of designing artificial sonar systems by mimicking natural, batlike solutions. Both ways have their virtues, but also their limitations. Approaching the problem from a purely engineering standpoint does not improve upon the reality of the biological system, nor does copying nature too closely always lead to the best results for artificial systems (think of early attempts at manned flight that, despite their efforts to imitate birds, or bats, never achieved success).

In order to make a possible connection between bat data and signal theory, one can imagine taking the place of a bat and trying to figure out "what it is like to be a bat" [159], evaluating the problems to be solved for both flying and hunting successfully by using only echolocation (in a first approximation), and trying to design optimal solutions for addressing the tasks at hand.

This approach justifies the introduction of the three phases that have been previously discussed for a hunting sequence. The first phase, "search," has the main objective of navigating in a given environment, using signals mostly for identifying and avoiding obstacles; the second phase, "approach," occurs as soon as a potential prey item has been spotted as a moving outlier in the learned environment. These are then followed by the third phase, "attack," in which the focus is on the confirmed prey until it is actually caught and eaten.

If we ignore the possible distractions caused by interference (from other bats hunting in the same area and emitting similar signals), as well as the benefit that could be gained from binaural processing, the simplest version of the problem becomes one of detection and estimation in the presence of noise: detection for identifying obstacles and prey items, and estimation for transforming time and frequency information into distance to target (via return time) and relative velocity (via Doppler shift), respectively.

Under the usual assumption that the returning echo is embedded in white Gaussian noise, the optimal receiver (derived from the maximization of the output signal-to-noise ratio or from a maximum likelihood argument) consists in *correlating* the observation with a replica of the emitted signal, i.e., in matched filtering. If we express the received echo as $r(t) = x(t - \tau) \exp\{i\xi t\} + n(t)$ in a first approximation, with $x(t)$ and $n(t)$ the emitted signal and the noise, and where τ and ξ stand for the time delay and the Doppler shift, the "signal part" at the output of the receiver is essentially the section at $\xi = 0$ of an *ambiguity function*, decorated with fluctuations due to the additive noise.

Detection is controlled by the output signal-to-noise ratio, which corresponds to the contrast between the *height* of the main peak in the envelope of the cross-correlation function (which turns out to be the *energy* E_x) and the level γ_0 of the background noise. As for *estimation* of time delay, the accuracy is directly related to the *width* of the "central" peak of the cross-correlation function, whose value depends on whether the receiver is fully coherent or not. A *coherent* receiver is sensitive to the *phase* of echoes and has a fine structure such that the width of the central peak in the cross-correlation function is approximately the reciprocal of the mean frequency. In contrast, a *semicoherent* receiver is insensitive to phase and only computes the envelope of the cross-correlation, whose width is approximately the reciprocal of the bandwidth.

Remark. In the case of wideband signals (i.e., when bandwidth and central frequency are of the same order of magnitude), the width of the envelope and of the central peak are about the same. This is the case for the bats we consider: to give an idea of what bats are capable of, a chirp sweeping from, say, 100 to 50 kHz permits a *range* accuracy of about 5 mm, when converting delay τ into distance d according to $d = ct$, with $c \approx 340$ m/s the sound velocity.

From an engineering point of view, a "good" signal for time delay estimation should combine constraints related to bandwidth and energy, both quantities being required to be large. This implies that time-limited tones are definitely ruled out since they would lead to a slowly-decaying correlation structure, with either a sharp central peak but confusing side peaks in a coherent scheme, or a large envelope width in a semi-coherent scheme. One could think of a very short transient which, by construction, would also have a highly peaked cross-correlation function and would thus permit a good time delay estimation. However, signal-to-noise ratio depends on energy, and the shorter the pulse, the larger the amplitude needed to attain a required value, a constraint which can find its limits in terms of physiology.

This brief analysis indicates, therefore, that it is advantageous to make use of chirps that can guarantee a large bandwidth, and which have a frequency sweep duration that determines the energy requirements for a given amplitude.

Remark. Estimation of relative velocity can be handled in a different way since, in a first approximation, it can be based on Doppler shift. In this context, "cross-correlation" could be conducted in the frequency domain, favoring long duration tones.

If we turn now to the bat solutions and go back to Figure 16.7, we can argue that the observed waveforms are in a very reasonable agreement with the engineering requirements. More precisely:

1.   "Search": In this phase, it is necessary for the bat to obtain information about both distances and relative velocities with respect to its environment. This explains the use of chirps with a large bandwidth-duration product for time delay; in addition, however, these signals contain an extension of almost constant frequency, which not only permits a fine estimation via Doppler shift, but also serves to increase the signal energy while maintaining the chirp rate (we will come back to this later).

2.   "Approach": This phase is supposed to start when contact has been established with a potential prey item. It is therefore important to keep in mind that the rest of the sequence will last for less than half a second. Considering that the bat flies at a speed of about 5 m/s, which is much higher than a mosquito's speed (on the order of 0.5 m/s), it becomes more important to accurately estimate delays rather than Doppler. Since the predator–prey distance is decreased, energy can be reduced and, in order to preserve performance in time, the tone component can be progressively decreased.

3.   "Attack": In this final phase, return times become shorter and shorter and the emission rate has to be increased accordingly. Assuming a sequential processing (no new emission before reception of the echo of the previous one), signal durations have to be reduced too. At some point, it might prove difficult to increase the chirp rate so as to sweep over the same frequency range. Since only bandwidth is important for a purpose of time delay estimation, a simpler solution is to shift the frequency spectrum of a "fundamental" chirp to a lower frequency support, and to introduce some nonlinear distortion, which creates a harmonic component to cover the upper part of the spectrum. One more advantage of this harmonic structure is that, assuming a coherent receiver, it allows for a better separation between the central peak and the first side peak, thus reducing ambiguity in the estimation (see Figure 16.8 for a toy model).

Doppler-tolerance revisited: The whole set of findings and interpretations reported on the previous pages strongly suggests that the use of chirps by bats could be the result of an optimal strategy based on a processing of echoes in the spirit of matched filtering. Even if we agree with this assumption, however, one question is still open: why do descending chirps have an upward concavity? A good answer was given by Richard A. Altes and Edward. L. Titlebaum in the early 1970s [160]: a *hyperbolic* decay of the instantaneous frequency (which, when taking a closer look at data, fits quite well

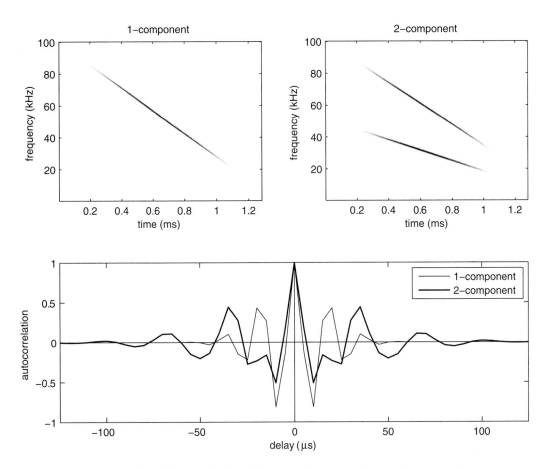

Figure 16.8 Disambiguation in time delay estimation. For a given bandwidth over the same duration, it proves more efficient to make use of a two-component chirp (top right) rather than a one-component chirp (top left), thanks to a better separation between the main peak and the side peak in the fine structure of the corresponding autocorrelation functions (bottom diagram). In both top diagrams, the time-frequency structure is displayed as a reassigned spectrogram, with a dynamic range of 30 dB.

with observations for *Myotis mystacinus* and, hence, for most common bats[2]) permits us to guarantee one more essential property in time delay estimation, namely *Doppler tolerance*.

Let us make this point more precise by first recalling that the concept of Doppler *shift* is only an approximation, and that it is mostly relevant for narrowband signals. In its most general form, Doppler affects waves by a *dilation* mechanism, according to which received waveforms are stretched/compressed by a factor $\eta = (c + v)/(c - v)$, with c the wave propagation celerity and v the relative velocity between the emitter and the

---

[2] See, e.g., www.ece.rice.edu/dsp/software/bat.shtml, from where it is possible to download an echolocation call of *Eptesicus fuscus* that has been commonly used as a benchmark signal in numerous papers on time-frequency analysis.

receiver. If $v \ll c$, a signal that is almost monochromatic therefore has its frequency ω_0, which is transformed into $\eta \omega_0 \approx \omega_0 + \delta\omega_0$, with $\delta\omega_0 \approx 2(v/c)\omega_0$ the Doppler shift. It thus turns out that the shift is frequency-dependent and, in the case of a wideband spectrum, one cannot consider that it applies equally to the whole bandwidth.

The second point to take into account is that estimations of delay and Doppler are coupled, which can be seen as a result of *uncertainty* between a pair of Fourier variables. In that respect, estimating a time delay by looking for the maximum of the correlation between signal and echo can be biased by the existence of a nonzero Doppler, which is *a priori* unknown. To overcome this difficulty, one can think of finding *Doppler-tolerant* waveforms such that time delay estimation would be unbiased, whatever the Doppler value. Writing down this problem in the wideband case for which Doppler corresponds to dilations (which is the bat situation, with frequency sweeps ranging from 100 kHz to 50 kHz, i.e., over one octave), we are led to an integral equation whose solution is given by hyperbolic chirps [160].

Besides a purely mathematical treatment, one can give a simple geometrical justification of this fact by combining time-frequency arguments. As for gravitational waves, matched filtering of a time-frequency localized chirp can be envisioned through a *path integration* along the instantaneous frequency trajectory of the emitted signal, which is considered to be a template. As for the echo, the Fourier transform of the dopplerized signal part $\sqrt{\eta}\, x(\eta t)$ is $X(\omega/\eta)/\sqrt{\eta}$. A pointwise application of this transformation along the instantaneous frequency $\omega(t)$ results in a modified trajectory such that

$$(t, \omega(t)) \rightarrow (\eta t, \omega(t)/\eta). \qquad (16.8)$$

Within this picture, Doppler tolerance just means an *invariant time-frequency trajectory* under such a transformation, for any η. It is easy to check that this is equivalent to:

$$\omega(\eta t) = \omega(t)/\eta \Rightarrow t\,\omega(t) = C \qquad (16.9)$$

or, in other words, to:

$$\omega(t) \propto 1/t, \qquad (16.10)$$

i.e., that the instantaneous frequency law should be hyperbolic.

Figure 16.9 illustrates graphically the optimality of hyperbolic chirps for Doppler tolerance, as compared to linear chirps sweeping over the same frequency band. In the linear case, the template trajectory and the dopplerized one have a much-reduced intersection as soon as Doppler is nonzero. In the hyperbolic case, on the other hand, the dopplerized trajectory *slides* along the template one, maximizing overlap in the path integration.

In the lab: Going beyond mere observations and assessing the actual existence of a (semi- or fully) coherent mechanism in the processing of echoes by bats has been the subject of intense investigations since the 1980s, in particular by James A. Simmons and his collaborators. Their approach was based on well-controlled indoor experiments. The idea was to train bats to answer a precise question in simple situations, to evaluate

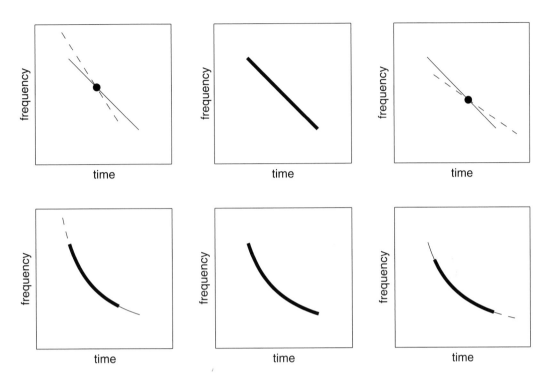

Figure 16.9 Doppler tolerance. Matched filtering can be interpreted in the time-frequency plane as a path integration along the instantaneous frequency trajectory. When Doppler is zero (middle column), signal (thin solid line) and echo (dotted line) trajectories coincide in both cases of linear (top row) and hyperbolic (bottom row) chirps. In case of a compression (left column) or of a stretching (right column), however, they differ. Intersection (thick solid line) essentially occurs in a point in the linear case, whereas it extends over a much longer support in the hyperbolic case, due to the *sliding* of the dotted line along the full line.

their performance limits, and to compare their results with predictions from matched filtering. A typical experiment was as follows. A bat was positioned on a Y-shaped platform in an anechoic room and was presented artificial echoes resulting from the sending of delayed replicas of emitted sounds via two loudspeakers that created the illusion of sounds coming from the left and from the right. Thanks to a reward protocol, the bat was trained to move in the direction of where "something" was to be identified, e.g., "where the closest echo was coming from." Once trained with easy separation situations, the performance could be tested by varying delays in a pseudo-random – yet controlled – fashion. In the event of a large enough separation, success was obtained with probability $P(\infty) = 1$ whereas, when the two virtual targets were at exactly the same distance, the probability was $P(0) = 1/2$. Repeating trials, it was possible to estimate an empirical success probability $P(\tau)$ as a function of the delay difference τ. Transforming this probability into a performance profile $g(\tau) = 2(1 - P(\tau))$, it turned out that $g(\tau)$ reproduced the envelope of the auto-correlation function of the emitted pulse

[161] fairly well. This remarkable result clearly favored the "semicoherent receiver" explanation for how bats estimate range.

Further experiments were then conducted in order to test the bat's performance in more complex tasks. In particular, "jitter" experiments were designed, in which the question (addressed to *Eptesicus fuscus*) was to detect delay *differences* in between echoes from consecutive pulses, simulating a fluttering target moving back and forth [162]. This proved this time a reproduction of the fine structure of the correlation function, suggesting the possible existence of a fully coherent receiver allowing for phase perception (see [163] for details and discussion). In parallel, different protocols questioned such a matched filtering mechanism: one can mention Bertel Møhl's experiments which, although based on Simmons' approach, sent a *time-reversed* replica of the bat's emitted pulse back to the bat [164]. Assuming matched filtering, the output of the receiver should no longer be a correlation, but a *convolution*. Convoluting a chirp by itself should result in a spreading, as opposed to the pulse compression attached to correlation. One should then expect a dramatic decrease in performance, but this was not actually observed to the degree that theoretical predictions from matched filtering had led the researchers to expect. This may suggest that matched filtering is not the sole mechanism in action, and that companion procedures aimed at making the receiver more robust (when departing from nominal conditions) could be involved. In particular, the previous brief analysis only considers successive echoes independently of each other, while more refined schemes could involve pulse-to-pulse processing in the spirit of *Motion Target Indication* (MTI) techniques that are in use in radar [165]. This would allow us to filter out echoes resulting from stationary obstacles and permit us to focus on moving targets, a situation that is not reproduced in the experiments of Simmons and Møhl, where the relative position of the virtual target with respect to the emitter is kept unchanged from one echolocation call to the next.

Built-in time-frequency receivers: Given some assumed processing, different options are offered for its implementation. When reduced to its simplest form, a matched filter is basically a *correlator* that compares an echo $e(t)$ with some copy $c(t)$ of the emitted signal by means of a standard inner product (in a L_2-sense):

$$r = \langle e, c \rangle = \int_{-\infty}^{\infty} e(t)\, c^*(t)\, \mathrm{d}t. \tag{16.11}$$

Equivalent formulations can therefore be obtained, either for the fully coherent output $r(t)$ or for the semicoherent one $|r(t)|^2$, by means of any alternative representation that preserves the (squared) inner product. Suitably chosen time-frequency representations offer such a possibility. For instance, it is known that:

$$r = \iint_{-\infty}^{\infty} F_e^{(h)}(t, \omega)\, F_c^{(h)*}(t, \omega)\, \mathrm{d}t \frac{\mathrm{d}\omega}{2\pi} \tag{16.12}$$

for any STFT with a window $h(t)$ of unit energy, while it follows from Moyal's formula (6.18) that

$$|r|^2 = \iint_{-\infty}^{\infty} W_e(t, \omega) W_c(t, \omega) \, dt \frac{d\omega}{2\pi} \qquad (16.13)$$

if the Wigner distribution is chosen as a time-frequency representation.

Both expressions amount to considering time-frequency signatures and, loosely speaking, to quantifying their possible overlap in the plane. However, they have distinctive interests with respect to echolocation:

- the linear case (16.12) may take advantage of the filterbank interpretation of a STFT, according to which

$$F_x^{(h)}(t, \omega) = \int_{-\infty}^{\infty} X(\xi) H^*(\xi - \omega) \, \exp\{i\xi t\} \frac{d\xi}{2\pi}. \qquad (16.14)$$

Whereas the classical, time-domain formulation favors a reading of a STFT as a collection of frequency spectra attached to different short-time locations, this alternative, yet equivalent, formulation expresses a STFT as the time evolution of the outputs of a collection of narrowband filters. This provides a starting point for the modeling of the tonotopic organization of the cochlea of most mammals, including bats. Indeed, acoustic waves impinging on the eardrum then propagate along the cochlea, with a spatial ordering of resonant frequencies that acts as a filterbank. As in (16.14), the evolution in time of the outputs of all those filters – which then go up to the auditory cortex – defines a time-frequency map that can be considered to physiologically encode the received signals. Making use of a STFT is of course an oversimplification, and refinements (in the spirit of the *mel-scale* in audio) can be envisioned, but the basic idea is simple: the external hearing system creates a natural time-frequency map whose units can be addressed for further processing.

- the quadratic case (16.13) is more interesting in a theoretical sense, since no physiological evidence can be produced to justify the choice of the Wigner distribution as a relevant time-frequency map. Nevertheless, the specific localization properties of the Wigner distribution make it a good starting point for a time-frequency correlator aimed at mimicking a semicoherent receiver. The same argument has already been used in the previous section devoted to gravitational waves, with the very same rationale that a chirp signal, whose time-frequency signature is essentially a time-frequency *trajectory*, can be detected via a path integration in the plane. One further point of interest is that such an interpretation paves the way for more robust procedures, in which the path integration can be enlarged to a ribbon: this may simply amount to smoothing the reference Wigner distribution, resulting in a variety of receivers that are discussed, e.g., in [7].

Giving bat echolocation processing a time-frequency formulation defines a whole field of research, of which we have only scratched the surface. Among the many possibilities that are offered, we will only mention that one way of reconciling the practical interest of the linear STFT and the theoretical interest of the quadratic Wigner-like distributions is to resort to spectrograms, which are simultaneously squared STFTs and smoothed Wigner distributions. This leads to optimal procedures of *spectrogram correlation* that have been extensively studied by Richard A. Altes [165–167] and further investigated – in the specific context of bat echolocation – by James A. Simmons and his collaborators [168]. We will not pursue this subject further here: instead, we will refer to those authors and their publications for more in-depth treatments.

16.3   Riemann-Like Special Functions

As announced in Chapter 2, we will now elaborate on a few examples stemming from mathematics to illustrate how time-frequency analysis can reveal inner structures in some special functions.

Zeta function: Let us start by coming back to the waveform $Z(t)$ defined in (2.3), which essentially describes Riemann's zeta function $\zeta(z)$ as a function of its imaginary part $\text{Im}\{z\} = t$ for the fixed value of its real part $\text{Re}\{z\} = \frac{1}{2}$. It has been shown that, although it is apparently highly erratic in the time domain (see Figure 2.5), this function has a well-organized structure in the time-frequency domain, made essentially of ascending chirps (see Figure 2.6).

This remarkable organization can be simply explained by resorting to the *Riemann-Siegel expansion* of $Z(t)$ [27]. When retaining its main sum, this expansion $Z_{RS}(t)$ represents $Z(t)$ as a series of oscillating terms, according to the expression:

$$Z_{RS}(t) = 2 \sum_{n=1}^{\lfloor \sqrt{t/2\pi} \rfloor} \frac{\cos\left(\theta(t) - t \log n\right)}{\sqrt{n}}, \tag{16.15}$$

with $\theta(t)$ as in (2.4), and where $\lfloor x \rfloor$ stands for the integer part of x.

The Riemann-Siegel expansion (16.15) therefore happens to be made of the superposition of time-varying modes, whose number increases as a function of time, and whose oscillations are controlled by the nonlinear phases $\Theta_n(t) = \theta(t) - t \log n$ for indices up to $n = \lfloor \sqrt{t/2\pi} \rfloor$. The "instantaneous frequencies" of such modes are given by $\omega_n(t) = \dot{\Theta}(t)$. Making use of the known large-t asymptotic form for $\theta(t)$ [27], those quantities can be approximated by

$$\omega_n(t) \approx \log\left(\frac{1}{n}\sqrt{t/2\pi}\right). \tag{16.16}$$

These "log-of-square-root" trajectories are precisely those uncovered by the time-frequency analysis reported in Figure 2.6. Figure 16.10 displays an amazing agreement between the model (16.16) and the actual localization of the energy ribbons.

Remark. For a sufficient time-frequency spacing between the modes, the same diagram with a reassigned spectrogram (which is not reproduced here) results in trajectories that are so highly localized that they are almost undistinguishable from the model. When the different modes get close, it is interesting to note that they create interference patterns taking the form of zeros aligned in between them. The overall constellation of those zeros is a clear illustration of the interference geometry discussed in Chapter 9.

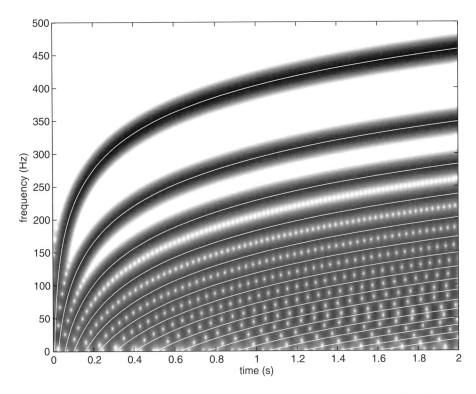

Figure 16.10 Zeta function 3. This figure reproduces Figure 2.6, with the Riemann-Siegel trajectories (16.16) (for $n = 1$–16, from top to bottom) superimposed as white lines. The energy scale is logarithmic with a dynamic range of 30 dB.

Psi function: As factorized by the Euler product (2.2), the zeta function is intimately related to prime numbers, whose distribution is also a fascinating object of interest *per se*. When properly sampled, this distribution can give rise to "signals" that can actually be *heard* [27], giving a new sense to the expression "music of primes," which is often used allegorically for referring to its supposed (secret) harmony.

In this respect, a special form of a prime counting function (i.e., a function aimed at quantifying how many primes are less than a given number) is known to offer an appealing connection with the zeros of the zeta function. This is (Riemann's) *psi function*, which relates to integer powers of primes as:

$$\psi(x) = \sum_{p^n < x} \log p, \tag{16.17}$$

where the summation is over all possible primes p and integers $n \in \mathbb{N}$.

This function can be advantageously decomposed into a smooth part and a fluctuating part $\tilde{\psi}(t)$ according to:

$$\psi(x) = x - \log 2\pi - \frac{1}{2} \log\left(1 - \frac{1}{x^2}\right) + \tilde{\psi}(x), \tag{16.18}$$

with

$$\tilde{\psi}(x) = -\sum_{n=1}^{\infty} \frac{x^{\frac{1}{2} \pm it_n}}{\frac{1}{2} \pm it_n}, \qquad (16.19)$$

and where the t_n's stand for the imaginary part of Riemann zeros with real part $\frac{1}{2}$.

Remark. Those zeros would account for *all* complex zeros of the zeta function if the "Riemann hypothesis" were true. Proving this hypothesis – i.e., stating that all non-trivial zeros z_n of the zeta function within the band defined by $0 < \mathrm{Re}\{z\} < 1$ lie along the critical line $\mathrm{Re}\{z\} = \frac{1}{2}$ – is one of the greatest unsolved problems in mathematics [28].

Focusing on the fluctuating part $\tilde{\psi}(t)$, a simple warping of the x-axis (together with a renormalization in amplitude) leads to the "signal":

$$S_\infty(t) = \exp\{-at/2\} \, \tilde{\psi}\left(\exp\{at\}\right), \qquad (16.20)$$

which can be viewed as $\lim_{N \to \infty} S_N(t)$, with

$$S_N(t) = -2\,\mathrm{Re}\left\{\sum_{n=1}^{N} \frac{\exp\{i\omega_n t\}}{\frac{1}{2} + it_n}\right\}, \qquad (16.21)$$

where $\omega_n = at_n$.

The primary interest of this formulation is to approximate the (warped) fluctuating part of the psi function by a finite superposition of *tones* whose frequencies are controlled by the Riemann zeros. Thanks to the free parameter a, it is possible to ensure that those tones are in the audio range [27], thus allowing for the playing of a "music of primes"... whose musical score can be written down thanks to a time-frequency analysis. This is what is reported in Figure 16.11.

In this example (adapted from [27]), the sampling rate has been fixed to 41.86 kHz so as to cover the audio frequency range up to $f_N = 20.93$ kHz. The lower frequency has been chosen as $f_1 = 27.5$ Hz, thus requiring us to have $a = 12.224$ and to make use of $N = 11{,}037$ terms in (16.21). This very large number of tones makes their spacing too small to be resolved as individual frequencies, giving rise to three different regimes that result from the interplay between the signal structure and the way it is analyzed:

1. In the first part of the signal (up to, say, half a second), the dense collection of tones in (16.21) maintains a sufficient phase coherence to make components in time appear highly localized exactly as the Dirac's δ-function results from the coherent summation of an infinite number of frequencies. Those time-frequency "vertical" lines are the signatures of the singularities, which reflect the "jumps" that occur in the representation (16.17) for all integer powers of the primes.

2. In the last part of the signal, more and more powers of primes are to be summed up, and they cannot be resolved in time. Combined with the fact that it remains impossible to resolve the corresponding individual frequencies, the overall structure resembles some kind of "white noise."

3. In the intermediate range, the different series of powers of primes tend to organize along exponential chirps. This can be understood by considering the counting function of zeros, which behaves as $N(t) \approx (t/2\pi) \log(t/2\pi e)$, resulting in a local increment between primes that is expressed as $\Delta t \approx t \exp\{-at\}$ in the warped representation (16.20) [27].

The visualization of those three regimes is naturally dependent on the chosen analysis window, which can or cannot resolve individual structures – be they tones, impulses or chirps – according to its time-frequency essential support. This argument will be invoked again and detailed further when analyzing the Weierstrass function.

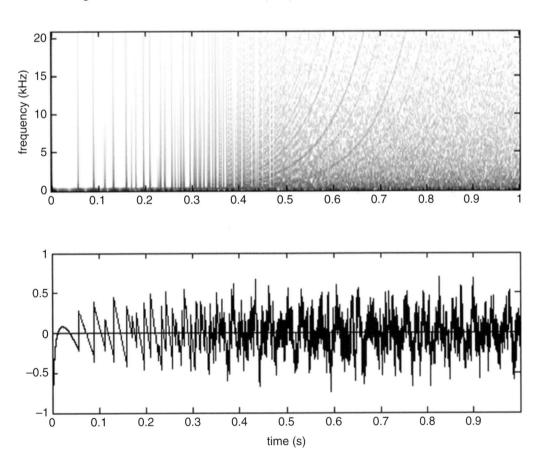

Figure 16.11 Music of primes. The bottom diagram displays the prime signal approximation $S_N(t)$, defined as the finite sum (16.21), with $N = 11{,}037$ terms. Its spectrogram is displayed in the top diagram, with a logarithmic energy scale over a dynamic range of 60 dB.

Sigma function: Besides the zeta function $\zeta(z)$ and the psi function $\psi(x)$ that have just been discussed, Bernhard Riemann defined a different *sigma function* $\sigma(t)$ as the following Fourier-type series:

$$\sigma(t) = \sum_{n=1}^{\infty} \frac{\sin \pi n^2 t}{n^2}. \tag{16.22}$$

For a long time, this 2-periodic function was thought to be simultaneously undifferentiable and continuous (it was even designed by Riemann for this purpose). It took almost a century before Joseph Gerver proved in 1969 that $\sigma(t)$ has a derivative at all points of the form $\{(2p+1)/(2q+1); p, q \in \mathbb{N}\}$, and only at those points. This result was later proved in a new way and refined by Stéphane Jaffard and Yves Meyer, who established – with wavelet methods [169] – that, in the vicinity of the differentiability points, the behavior of $\sigma(t)$ was controlled by *oscillating singularities*, i.e., a specific form of *power-law chirps*, which enter the general model:

$$C_{\alpha,\beta,\gamma}(t) = |t|^{\alpha} \exp\{i\pi \gamma |t|^{-\beta}\}, \tag{16.23}$$

with α and $\beta(> 0)$ the chirp exponents, and γ the chirp rate.

Remark. It would seem that, with a phase term $\varphi = 0$, the gravitational wave chirp (16.1) could be written as the anticausal (with respect to the coalescence time t_0) real part of a power-law chirp (16.23), with parameters $\alpha = -1/4$, $\beta = -5/8$ and $\gamma = 2d$. This is, however, only formal and somewhat hazardous since a proper definition of (16.23) requires $\beta > 0$ in order to satisfy the slowly-varying condition for a chirp, namely $|\ddot{\varphi}(t)/\dot{\varphi}(t)|^2 \ll 1$ if $\varphi(t)$ stands for the phase. Formal power-law chirps with $\beta < 0$ may behave as proper ones on a domain not too close to the singularity, but they fail when approaching it. Similarly, if we admit that the instantaneous frequency of the chirp (16.23) varies as $|t|^{-\beta-1}$, one could be tempted to make use of (16.23) with $\beta = 0$ for modeling the Doppler-tolerant bat chirps whose instantaneous frequency is, according to (16.10), precisely hyperbolic. This is not directly possible within the framework of (16.23), but requires a logarithmic phase in place of a power-law one. One can note, however, that – defined this way – hyperbolic chirps are valid chirps everywhere.

It turns out [169] that, in the vicinity of $t = 1$, the graph of the Riemann function $\sigma(1 + \tau)$ consists of a straight line of negative slope $-\pi/2$, decorated with an infinite sum over $n \geq 1$ of (weighted) power-law chirps $C_{3/2,1,n^2}(\tau)$. We thus have

$$\sigma(1 + \tau) = \sigma(1) - \pi\tau/2 + O\left(|\tau|^{3/2}\right) \tag{16.24}$$

when $\tau \to 0$. Proved this way, differentiability appears to be accompanied by oscillating singularities whose structure is revealed in a particularly appealing manner by a time-frequency analysis, as shown in Figure 16.12.

Weierstrass function: A different form of undifferentiable, though continuous, function was introduced in 1872 by Karl Weierstrass. In its original formulation, it was

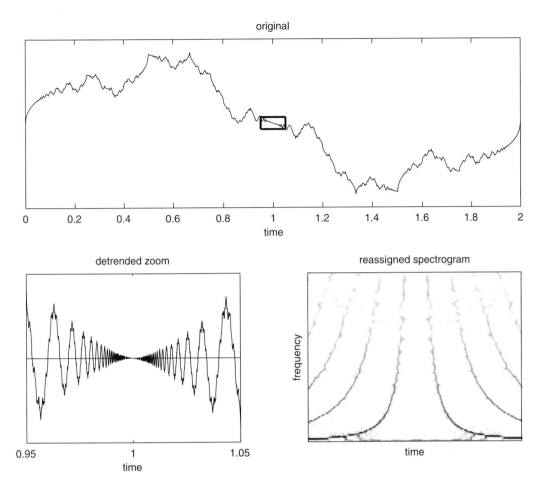

Figure 16.12 Riemann's sigma function. Top : graph of the function over the fundamental period [0, 2]. Bottom left: enlargement of the function within the box delineated by a thick line in the top diagram, after detrending. Bottom right: reassigned spectrogram of the (detrended) enlarged detail, with a dynamic range of 30 dB. © 2001 Society of Photo-Optical Instrumentation Engineers (SPIE). Adapted, with permission, from [32].

simply expressed as a kind of Fourier series with *geometrically spaced* modes, namely:

$$\tilde{W}_{H,\lambda}(t) = \sum_{n=0}^{\infty} \lambda^{-nH} \cos \lambda^n t, \qquad (16.25)$$

with $\lambda > 1$ and $0 < H < 1$. The nondifferentiability of $\tilde{W}_{H,\lambda}(t)$ ends up with a *fractal* structure [170] for the graph of such a function, with noninteger dimension $2 - H$. Although fractal, $\tilde{W}_{H,\lambda}(t)$ is not exactly self-similar, due to the summation that begins with the index $n = 0$ and thus creates a low-frequency cutoff. This limitation prompted Benoît Mandelbrot to slightly modify the original definition so as to guarantee a more complete scale invariance. The new definition reads [171]:

$$W_{H,\lambda}(t) = \sum_{n=-\infty}^{\infty} \lambda^{-nH}\left(1 - \exp\{i\,\lambda^n t\}\right),$$ (16.26)

and it is easy to check that it is such that $W_{H,\lambda}(\lambda^k t) = \lambda^{kH} W_{H,\lambda}(t)$ for any $k \in \mathbb{Z}$. This is the signature of a *discrete scale invariance* [39]: for a proper renormalization in amplitude, the function is identical to any of its versions in time after a rescaling with respect to an integer power of the preferred scale ratio λ.

Remark. Mandelbrot's definition can be further generalized in at least two ways. First, by introducing a possibly random phase term in the summation, for the sake of modeling *statistically self-similar* stochastic processes. Second, by replacing the complex exponential with any periodic function that is continuously differentiable at $t = 0$. We will not consider such variations here; interested readers are referred to [171] and [172] for further developments.

By construction, the Weierstrass-Mandelbrot function $W_{H,\lambda}(t)$ is made of a series of tones. At the same time, it is scale-invariant, which suggests that it could also be expanded by means of self-similar functions. As developed in [172], this is indeed the case and the key for this alternate representation can be found in a variation on the Fourier transform that is naturally built upon scale-invariant building blocks, namely on the *hyperbolic chirps*:

$$c_{H,s}(t) = t^{H+is}\,\mathbf{1}_{]0,\infty)}(t).$$ (16.27)

This variation is the so-called *Mellin transform* [173]. Considering causal signals defined for $t \geq 0$ and adapting to the above form of (16.27), we will now adopt the following definition for the Mellin transform:

$$(\mathbf{M}_H x)(s) = \int_0^{+\infty} x(t)\,c_{H,s}^*(t)\,\frac{dt}{t^{2H+1}},$$ (16.28)

with the corresponding inversion formula:

$$x(t) = \int_{-\infty}^{\infty} (\mathbf{M}_H x)(s)\,c_{H,s}(t)\,\frac{ds}{2\pi}.$$ (16.29)

Remark. According to (16.27), the "instantaneous frequency" of the hyperbolic Mellin chirps is given by $\omega(t) = s/t$, and the s parameter therefore has the status of a hyperbolic chirp rate. Whereas the Fourier transform is based on tones, the Mellin transform is based on hyperbolic chirps, the latter being derivable from the former by an appropriate time warping. More precisely, if we denote the Fourier transform by $\mathbf{F}x$ and if we introduce the warped signal $x_H(t) = \exp\{-Ht\}\,x(e^t)$, we can formally write $\mathbf{M}_H x = \mathbf{F}x_H$.

Applying the Mellin machinery to the Weierstrass-Mandelbrot function (16.26), we readily obtain the desired expansion [171, 172]:

$$W_{H,\lambda}(t) = -\frac{1}{\log \lambda} \sum_{m=-\infty}^{\infty} \exp\{-i\pi\beta_m/2\}\,\Gamma(-\beta_m)\,c_{H,m/\log \lambda}(t),$$ (16.30)

with $\beta_m = H + im/\log \lambda$, and $\Gamma(.)$ the Gamma function.

Under its form (16.30), the real part of the Weierstrass-Mandelbrot function is made of *oscillating contributions* due to the terms with indexes $m \neq 0$, superimposed to a *low-frequency trend* $T_{H,\lambda}(t)$, which is captured by $m = 0$ and reads:

$$T_{H,\lambda}(t) = \frac{\Gamma(1-H)\,\cos(\pi H/2)}{H \log \lambda}\, t^H. \tag{16.31}$$

Figure 16.13 displays an example of the Weierstrass-Mandelbrot function and its detrended version. A time-frequency plot of the latter reveals a *mixed structure* that tends to favor chirps at low frequencies and tones at high frequencies. This can be given a simple time-frequency interpretation by considering in parallel the two structures

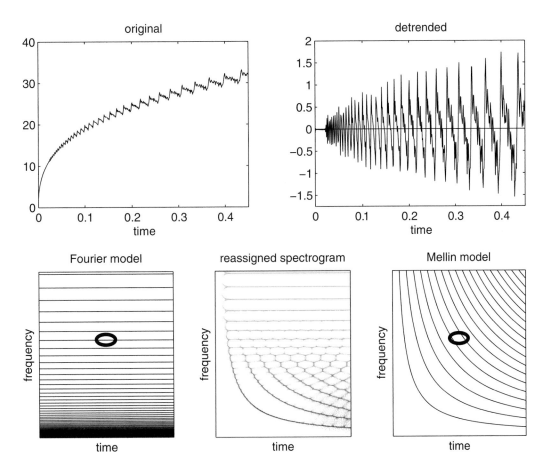

Figure 16.13 Weierstrass function. Top left: real part of the Weierstrass-Mandelbrot function $W_{H,\lambda}(t)$ over the time interval $[0, 1/2]$, with parameters $H = 0.3$ and $\lambda = 1.09$. Top right: detrended version of $W_{H,\lambda}(t)$. Bottom row: the reassigned spectrogram (middle) of the detrended version of $W_{H,\lambda}(t)$ is plotted in between the time-frequency structures of the tone model (16.26) (left) and of the chirp model (16.30) (right). In both diagrams, the thick line ellipse is a symbolic representation of the time-frequency domain occupied by the Gaussian short-time window used to compute the reassigned spectrogram.

attached to the tone and chirp models, and by putting them in perspective with the time-frequency extent of the analyzing window. What happens is in fact the following: when analyzing frequencies that are high enough, the separation between tones is sufficiently large to enable us to "see" each tone individually through the time-frequency window of the analysis, something that cannot be achieved with chirps. As a result, the computed time-frequency distribution emphasizes the tone structure (tones that actually result from an interference process between chirps). The situation is reversed when analyzing low frequencies. In this case, chirps are "seen" individually, while many tones enter the time-frequency analysis window. Chirps appear as the result of interfering tones, and the computed time-frequency distribution emphasizes the chirp structure.

17 Conclusion

And now, our journey comes to an end. Although we did explore a number of aspects of time-frequency analysis in this book, we merely scratched the surface in terms of all the possibilities.

Although this book does emphasize 1D signals and time series, with the goal of explaining the basics and various interpretations, this author hopes that this will serve as an entry point for further study in, for example, bivariate signals [174], or higher-dimensional data, even possibly attached to non-Euclidean structures [176].

As pointed out in the Introduction, the deluge of data that we experience daily in today's era of "big data" does not preclude thinking of signals on a smaller scale; we continue to study the fine structures of signals, which still perform the same functions in learning approaches based on large raw datasets. Indeed, this focus on the fine structures of signals is reminiscent of the situation we face in physics, where going down to smaller and smaller scales means entering a world with constraints and limitations that call for dedicated tools and appropriate interpretations. In the context of time-frequency analysis, this pertains essentially to the "uncertainty" that prevents a pointwise joint localization of any signal, and that echoes what happens in quantum mechanics (with position and momentum in place of time and frequency). We have not commented much about the connections that exist between the two domains (this can be found, e.g., in [22] or [7], and references therein), but such connections do exist, at least in terms of formalisms. One interesting point is that this allows us to go back and forth between the two fields and, crossing references from complementary literatures, to import physics tools into signal theory [22], and vice versa [175].

The uncertainty limits faced by time-frequency analysis show that thinking of this subject in classical terms is not really the best way to approach it. Loosely speaking, there is no way of representing a signal jointly in time and frequency so that every property known to hold for a time-independent spectrum (and its transformations) would be preserved *mutatis mutandis* when making a spectrum time-dependent. If we acknowledge this fact, we have two options: the first one is to stick to a "classical" perspective that discourages studying anything that does not fit "inside the box," or that cannot be neatly resolved. It is worth noting that, historically, this attitude prevailed in the signal processing community for some time, and no mainstream studies on the subject were done until the early 1980s; even then, time-frequency analysis was looked on with some skepticism. In reference to Sir Winston Churchill's famous quote "*A pessimist sees the difficulty in every opportunity; an optimist sees the opportunity in every difficulty*," the other option is more optimistic. It amounts to accepting the ultimate time-frequency limits fixed by uncertainty (in fact, by Fourier analysis, as explained in Chapter 5) and reconsidering the possible unicity of a time-frequency representation in "nonclassical" terms, i.e., as subject to user-based constraints. On the one hand, this has the conceptual flavor of a *regularization*, analogous to the successful approach that is commonly used with, e.g., ill-posed inverse problems. On the other hand, this change of viewpoint does also present the advantage of placing the user in a more central position, with the possibility of tailoring specific solutions to specific needs.

> This book has explored some of the many tracks that shape the time-frequency landscape. While it outlined territories that are now well known, it also pointed out some more remote areas that are worthy of further exploration, as well as some summits that remain to be conquered. As with any field of research, the field of time-frequency analysis presents a horizon that not only seems unattainable, but that also continues to expand with every step we take in its direction.

18 Annex: Software Tools

The primary purpose of this book was to develop general concepts for time-frequency analysis and to focus on interpretations. To this end, signals and processes have been considered throughout the text in a continuous-time formalism, while in actual practice, they of course most often take a discrete-time form, due either to sampling of continuous waveforms or to their nature as time series. Although not explicitly mentioned, all figures in this volume have been produced in a discrete-time setting, thanks to a number of software tools that are publicly available. More precisely, the core of most of the basic time-frequency methods discussed in this book is contained in a "Time-Frequency Tool-Box" (TFTB) whose Matlab implementation – initially developed by François Auger, Paulo Gonçalves, Olivier Lemoine, and the author, within the framework of a CNRS project supported by GdR TdSI – can be downloaded from

http://tftb.nongnu.org

A Python version of the same – developed by Jaidev Deshpande – can be found at

https://github.com/scikit-signal/pytftb.

Building upon this TFTB toolbox, some specific extensions related to multitapering, stationarity tests (within the ANR "StaRAC" project (ANR-07-BLAN-0191)), and sparsity have benefited from contributions of Pierre Borgnat, Paul Honeine, Nelly Pustelnik, Cédric Richard, and Jun Xiao: these can be found at

http://perso.ens-lyon.fr/patrick.flandrin/software2.html.

More recent developments have been achieved by the ANR "ASTRES" project (ANR-13-BS03-0002-01), with additional contributions from François Auger, Philippe Depalle, Dominique Fourer, Jinane Harmouche, Thomas Oberlin, Sylvain Meignen, and Jeremy Schmitt. The corresponding Matlab implementations can be found at

https://github.com/dfourer/ASTRES_toolbox.

As for Empirical Mode Decomposition, the basic algorithm – developed by Gabriel Rilling and the author – is downloadable from

http://perso.ens-lyon.fr/patrick.flandrin/emd.html

in Matlab and C++ forms, while a Python version – developed by Jaidev Deshpande – can be found at

https://github.com/jaidevd/pyhht.

This Annex only mentions time-frequency toolkits that have been used throughout this book, but many more such toolkits can be found online. This is particularly true when venturing into the related field of wavelets; for more on this subject, interested readers are referred to works such as [20] and the references therein.

References

[1] É.-L. Scott, "Principes de phonautographie" (sealed manuscript), Paris: Archives de l'Académie des sciences, January 26, 1857.

[2] É.-L. Scott, "Inscription automatique des sons de l'air au moyen d'une oreille artificielle," *Comptes Rendus Ac. Sc. Paris*, Vol. LIII, No. 3, pp. 108–11, 1861.

[3] W. Koenig, H. K. Dunn, and D. Y. Lacy, "The sound spectrograph," *J. Acoust. Soc. Amer.*, Vol. 18, No. 1, pp. 19–49, 1946.

[4] R. K. Potter, G. A. Kopp, and H. C. Green, *Visible Speech*, Amsterdam: D. von Nostrand, 1947.

[5] B. P. Abbott et al., "Observation of gravitational waves from a binary black hole merger," *Phys. Rev. Lett.*, Vol. 116, No. 6, pp. 061102-1–061102-16, 2016.

[6] É. Chassande-Mottin, S. Jaffard, and Y. Meyer, "Des ondelettes pour détecter les ondes gravitationnelles," *Gazette de la Société Mathématique de France*, No. 148, pp. 61–4, 2016.

[7] P. Flandrin, *Time-Frequency/Time-Scale Analysis*, San Diego, CA: Academic Press, 1999.

[8] D. Donoho, "50 years of Data Science," *J. Comput. Graph. Stat.*, Vol. 26, No. 4, pp. 745–66, 2017.

[9] P. A. R. Ade et al., "*Planck* 2013 results. I. Overview of products and scientific results," *Astron. Astrophys.*, Vol. 571, A1, 2014.

[10] N. Wiener, *Extrapolation, Interpolation, and Smoothing of Stationary Time Series*, New York: John Wiley & Sons, 1949.

[11] A. Blanc-Lapierre and R. Fortet, *Théorie des Fonctions Aléatoires*, Paris: Masson, 1953.

[12] J. Fourier, *Théorie Analytique de la Chaleur*, Paris: Firmon-Didot, 1822.

[13] B. Escudié, C. Gazanhes, H. Tachoire, and V. Torra, *Des Cordes aux Ondelettes*, Aix-en-Provence: Publications de l'Université de Provence, 2001.

[14] J. W. Cooley and J. W. Tukey, "An algorithm for the machine calculation of complex Fourier series," *Math. Comput.*, Vol. 19, pp. 297–301, 1965.

[15] J. P. Kahane, "Fourier, un mathématicien inspiré par la physique," *Images de la Physique 2009*, Article 01, CNRS, 2009. www.cnrs.fr/publications/imagesdelaphysique/couv-PDF/IdP2009/Article_01.pdf

[16] N. Wiener, *Ex-Prodigy: My Childhood and Youth*, Cambridge, MA: MIT Press, 1964.

[17] A. Grossmann and J. Morlet, "Decomposition of Hardy functions into square-integrable wavelets of constant shape," *SIAM J. Math. Anal.*, Vol. 15, pp. 723–36, 1984.

[18] Y. Meyer, *Ondelettes et Opérateurs I. Ondelettes*, Paris: Hermann, 1990.

[19] I. Daubechies, *Ten Lectures on Wavelets*, Philadelphia, PA: SIAM, 1992.

[20] S. Mallat, *A Wavelet Tour of Signal Processing – The Sparse Way* (3rd edn.), Burlington, MA: Academic Press, 2009.

[21] B. Boashash (ed.), *Time-Frequency Signal Analysis and Processing – A Comprehensive Reference* (2nd edn.), Amsterdam: Elsevier, 2016.

[22] L. Cohen, *Time-Frequency Analysis*, Englewood Cliffs, NJ: Prentice Hall, 1995.

[23] K. Gröchenig, *Foundations of Time-Frequency Analysis*, Boston, MA: Birkhäuser, 2011.

[24] L. Spallanzani, *Lettere sopra il sospetto di un nuovo senso nei pipistrelli*, Torino: Nella Stamperia reale, 1794.

[25] G. W. Pierce and D. R. Griffin, "Experimental determination of supersonic notes emitted by bats," *J. Mammal.*, Vol. 19, pp. 454–5, 1938.

[26] J. A. Simmons, "A view of the world through the bat's ear: The formation of acoustic images in echolocation," *Cognition*, Vol. 33, pp. 55–199, 1989.

[27] M. V. Berry, "Hearing the music of the primes: Auditory complementarity and the siren song of zeta," *J. Phys. A: Math. Theor.*, Vol. 45, 382001, 2012.

[28] B. Mazur and W. Stein, *Prime Numbers and the Riemann Hypothesis*, Cambridge: Cambridge University Press, 2016.

[29] D. Gabor, "Theory of communication," *J. IEE*, Vol. 93, pp. 429–41, 1946.

[30] J. Ville, "Théorie et applications de la notion de signal analytique," *Câbles et Transmissions*, Vol. 2A, pp. 61–74, 1948.

[31] É. Chassande-Mottin and P. Flandrin, "On the stationary phase approximation of chirp spectra," in *Proceedings IEEE International Symposium on Time-Frequency and Time-Scale Analysis*, pp. 117–20, Pittsburgh, PA: IEEE, 1998.

[32] P. Flandrin, "Time frequency and chirps," in *Proceedings of SPIE 4391, Wavelet Applications VIII*, 161, 2001.

[33] D. Vakman, "On the analytic signal, the Teager-Kaiser energy algorithm, and other methods for defining amplitude and frequency," *IEEE Trans. Signal Proc.*, Vol. 44, No. 4, pp. 791–7, 1996.

[34] P. Flandrin, J. Sageloli, J. P. Sessarego, and M. Zakharia, "Application of time-frequency analysis to the characterization of surface waves on elastic targets," *Acoust. Lett.*, Vol. 10, No. 2, pp. 23–8, 1986.

[35] L. R. O. Storey, "An investigation of whistling atmospherics," *Proc. Phil. Trans. Roy. Soc. A*, Vol. 246, No. 908, pp. 113–41, 1953.

[36] N. K.-R. Kevlahan and J. C. Vassilicos, "The space and scale dependencies of the self-similar structure of turbulence," *Proc. Phil. Trans. Roy. Soc. A*, Vol. 447, No. 1930, pp. 341–63, 1994.

[37] S. J. Schiff et al., "Brain chirps: Spectrographic signatures of epileptic seizures," *Clin. Neurophysiol.*, Vol. 111, No. 6, pp. 953–8, 2000.

[38] J. Duchêne, D. Devedeux, S. Mansour, and C. Marque, "Analyzing uterine EMG: Tracking instantaneous burst frequency," *IEEE Eng. Med. Biol. Mag.* Vol. 14, No. 2, pp. 125–32, 1995.

[39] D. Sornette, "Discrete scale invariance and complex dimensions," *Phys. Rep.*, Vol. 297, No. 5, pp. 239–70, 1998.

[40] J. Simmonds and D. MacLennan, *Fisheries Acoustics: Theory and Practice* (2nd edn.), Oxford: Blackwell Science, 2005.

[41] P. L. Goupillaud, "Signal design in the 'vibroseis'® technique," *Geophysics*, Vol. 41, pp. 1291–304, 1976.

[42] A. Papoulis, *Probability, Random Variables, and Stochastic Processes*, New York: McGraw-Hill, 1965.

[43] M. Vetterli, J. Kovačević, and V. K. Goyal, *Foundations of Signal Processing*, Cambridge: Cambridge University Press, 2014.

[44] M. Unser and P. Tafti, *An Introduction to Sparse Stochastic Processes*, Cambridge: Cambridge University Press, 2014.

[45] J. Schreier and L. L. Scharf, *Statistical Signal Processing of Complex-Valued Data: The Theory of Improper and Noncircular Signals*, Cambridge: Cambridge University Press, 2010.

[46] E. L. Pugh, "The generalized analytic signal," *J. Math. Anal. Appl.*, Vol. 89, No. 2, pp. 674–99, 1982.

[47] M. T. Heideman, D. H. Johnson, and C. S. Burrus, "Gauss and the history of the Fast Fourier Transform," *IEEE Acoust. Speech Signal Proc. Mag.*, Vol. 1, No. 4, pp. 14–21, 1984.

[48] N. G. De Bruijn, "Uncertainty principles in Fourier analysis," in *Inequalities* (O. Shisha, ed.), pp. 57–71, New York: Academic Press, 1967.

[49] R. J. Glauber, "Coherent and incoherent states of the radiation field," *Phys. Rev.*, Vol. 131, pp. 2766–88, 1963.

[50] C. W. Helström, "An expansion of a signal in Gaussian elementary signals," *IEEE Trans. Inf. Theory*, Vol. 12, No. 1, pp. 81–2, 1968.

[51] K. Husimi, "Some formal properties of the density matrix," *Proc. Phys. Math. Soc. Jpn.*, Vol. 22, pp. 264–314, 1940.

[52] W. Heisenberg, "Über den anschaulichen Inhalt der quantentheoretischen Kinematic und Mechanik," *Zeit. Physik*, Vol. 43, No. 3–4, pp. 172–98, 1927.

[53] H. Weyl, *Gruppentheorie und Quantenmechanik*, Leipzig: S. Hirzel, 1928.

[54] E. Schrödinger, "Zum Heisenbergschen Unschärfeprinzip," *Sitzungsberichte der Preussischen Akademie der Wissenschaften*, Vol. 14, pp. 296–303, 1930. English translation available at http://arxiv.org/abs/quant-ph/9903100.

[55] I. I. Hirschman, "A note on entropy," *Amer. J. Math.*, Vol. 79, No. 1, pp. 152–6, 1957.

[56] W. Beckner, "Inequalities in Fourier analysis," *Ann. Math.*, Vol. 102, No. 1, pp. 159–82, 1975.

[57] G. B. Folland and A. Sitaram, "The uncertainty principle: A mathematical survey," *J. Fourier Anal. Appl.*, Vol. 3, No. 3, pp. 207–38, 1997.

[58] A. Dembo, T. M. Cover, and J. A. Thomas, "Information theoretic inequalities," *IEEE Trans. Inf. Theory*, Vol. 37, No. 6, pp. 1501–18, 1991.

[59] P. Flandrin, "The many roads to time-frequency," in *Nonlinear and Nonstationary Signal Processing* (A. Walden et al., eds.), Isaac Newton Institute Series, pp. 275–91, Cambridge: Cambridge University Press, 2001.

[60] P. M. Woodward, *Probability and Information Theory with Applications to Radar*, London: Pergamon Press, 1953.

[61] P. Flandrin, "Ambiguity functions," in [21], Article 5.1, pp. 238–43.

[62] G. L. Turin, "An introduction to matched filters," *IRE Trans. Inf. Theory*, Vol. 6, No. 3, pp. 311–29, 1960.

[63] E. P. Wigner, "On the quantum correction for thermodynamic equilibrium," *Phys. Rev.*, Vol. 40, pp. 749–59, 1932.

[64] R. L. Hudson, "When is the Wigner quasi-probabilty density non-negative?," *Rep. Math. Phys.*, Vol. 6, No. 2, pp. 249–52, 1974.

[65] O. Rioul and P. Flandrin, "Time-scale energy distributions: A general class extending wavelet transforms," *IEEE Trans. Acoust. Speech Signal Proc.*, Vol. 40, No. 7, pp. 1746–57, 1992.

[66] D. B. Percival and A. T. Walden, *Spectral Analysis for Physical Applications: Multitaper and Conventional Univariate Techniques*, Cambridge: Cambridge University Press, 1993.

[67] G. B. Folland, *Harmonic Analysis in Phase Space*, Ann. of Math. Studies No. 122, Princeton, NJ: Princeton University Press, 1989.

[68] T. A. C. M. Claasen and W. F. G. Mecklenbräuker, "The Wigner distribution – A tool for time-frequency signal analysis. Part I: Continuous-time signals, Part II: Discrete-time signals, Part III: Relations with other time-frequency signal transformations," *Philips J. Res.*, Vol. 35, pp. 217–50, 276–300, and 372–89, 1980.

[69] W. F. G. Mecklenbräuker and F. Hlawatsch (eds.), *The Wigner Distribution: Theory and Applications in Signal Processing*, Amsterdam: Elsevier, 1998.

[70] E. H. Lieb, "Integral bounds for radar ambiguity functions and Wigner distributions," *J. Math. Phys.*, Vol. 31, pp. 594–9, 1990.

[71] R. G. Baraniuk, P. Flandrin, A. J. E. M. Janssen, and O. J. J. Michel, "Measuring time-frequency information content using the Rényi entropies," *IEEE Trans. Inf. Theory*, Vol. 47, No. 4, pp. 1391–409, 2001.

[72] P. Flandrin, "Maximum signal energy concentration in a time-frequency domain," in *Proceedings of IEEE International Conference on Acoustics, Speech and Signal Processing ICASSP-88*, pp. 2176–79, New York, 1988.

[73] A. J. E. M. Janssen, "Optimality property of the Gaussian window spectrogram," *IEEE Trans. Signal Proc.*, Vol. 39, pp. 202–4, 1991.

[74] P. Flandrin, "Cross-terms and localization in time-frequency energy distributions," in [21], Article 4.2, pp. 151–8.

[75] R. A. Silverman, "Locally stationary random processes," *IEEE Trans. Inf. Theory*, Vol. 3, pp. 182–7, 1957.

[76] P. Borgnat, P. Flandrin, P. Honeine, C. Richard, and J. Xiao, "Testing stationarity with surrogates: A time-frequency approach," *IEEE Trans. Signal Proc.*, Vol. 58, No. 7, pp. 3459–70, 2010.

[77] B. B. Mandelbrot, *The Fractal Geometry of Nature* (3rd edn.), New York: W. H. Freeman and Co., 1983.

[78] J. Makhoul, "Linear prediction: A tutorial review," *Proc. IEEE*, Vol. 63, No. 4, pp. 561–80.

[79] P. Abry, P. Gonçalvès, and P. Flandrin, "Wavelets, spectrum analysis, and $1/f$ processes," in *Wavelets and Statistics* (A. Antoniadis et al., eds.), *Lecture Notes in Statistics*, Vol. 103, pp. 15–29, New York: Springer-Verlag, 1995.

[80] J. Theiler, S. Eubank, A. Longtin, B. Galdrikian, and J. D. Farmer, "Testing for nonlinearity in time series: The method of surrogate data," *Physica D*, Vol. 58, No. 1–4, pp. 77–94, 1992.

[81] M. Basseville, "Distances measures for signal processing and pattern recognition," *Signal Proc.*, Vol. 18, No. 4, pp. 349–69, 1989.

[82] F. Hlawatsch and P. Flandrin, "The interference structure of the Wigner distribution and related time-frequency signal representations," in [69], pp. 59–133.

[83] A. Okabe, B. Boots, K. Sugihara, and S. N. Chiu, *Spatial Tessellations: Concepts and Applications of Voronoi Diagrams* (2nd edn.), New York: John Wiley & Sons, 2000.

[84] P. Flandrin and P. Gonçalvès, "Geometry of affine time-frequency distributions," *Appl. Comp. Harm. Anal.*, Vol. 3, No. 1, pp. 10–39, 1996.

[85] K. Kodera, C. de Villedary, and R. Gendrin, "A new method for the numerical analysis of non-stationary signals," *Phys. Earth Plan. Int.*, Vol. 12, pp. 142–50, 1976.

[86] K. Kodera, R. Gendrin, and C. de Villedary, "Analysis of time-varying signals with small *BT* values," *IEEE Trans. Acoust. Speech Signal Proc.*, Vol. 26, pp. 64–76, 1978.

[87] F. Auger and P. Flandrin, "Improving the readability of time-frequency and time-scale representations by the reassignment method," *IEEE Trans. Acoust. Speech Signal Proc.*, Vol. 43, No. 5, pp. 1068–89, 1995.

[88] P. Flandrin, F. Auger, and É. Chassande-Mottin, "Time-frequency reassignment – From principles to algorithms," in *Applications in Time-Frequency Signal Processing* (A. Papandreou-Suppappola, ed.), Chapter 5, pp. 179–203, Boca Raton, FL: CRC Press, 2003.

[89] S. A. Fulop and K. Fitz, "A spectrogram for the twenty-first century," *Acoust. Today*, Vol. 2, No. 3, pp. 26–33, 2006.

[90] P. D. Welch, "The use of the Fast Fourier Transform for the estimation of power spectra: A method based on time averaging over short, modified periodograms," *IEEE Trans. Audio Electroacoust.*, Vol. AU-15, pp. 70–3, 1967.

[91] D. J. Thomson, "Spectrum estimation and harmonic analysis," *Proc. IEEE*, Vol. 70, No. 9, pp. 1055–96, 1982.

[92] J. Xiao and P. Flandrin, "Multitaper time-frequency reassignment for nonstationary spectrum estimation and chirp enhancement," *IEEE Trans. Signal Proc.*, Vol. 55, No. 6, pp. 2851–60, 2007.

[93] I. Daubechies and S. Maes, "A nonlinear squeezing of the continuous wavelet transform based on auditory nerve models," in *Wavelets in Medicine and Biology* (A. Aldroubi and M. Unser, eds.), pp. 527–46, Boca Raton, FL: CRC Press, 1996.

[94] F. Auger, P. Flandrin, Y. T. Lin, S. McLaughlin, S. Meignen, T. Oberlin, and H. T. Wu, "Time-frequency reassignment and synchrosqueezing," *IEEE Signal Proc. Mag.*, Vol. 30, No. 6, pp. 32–41, 2013.

[95] P. Flandrin, "A note on reassigned Gabor spectrograms of Hermite functions," *J. Fourier Anal. Appl.*, Vol. 19, No. 2, pp. 285–95, 2012.

[96] P. Flandrin, "Some features of time-frequency representations of multicomponent signals," in *Proceedings of IEEE International Conference on Acoustics, Speech and Signal Processing ICASSP-84*, pp. 41B.4.1–41B.4.4, San Diego, CA, 1984.

[97] D. L. Donoho, "Compressed sensing," *IEEE Trans. Inf. Theory*, Vol. 52, No. 4, pp. 1289–306, 2006.

[98] E. Candès, J. Romberg, and T. Tao, "Stable signal recovery from incomplete and inaccurate measurements," *Comm. Pure Appl. Math.*, Vol. 59, No. 8, pp. 1207–23, 2006.

[99] P. Flandrin and P. Borgnat, "Time-frequency energy distributions meet compressed sensing," *IEEE Trans. Signal Proc.*, Vol. 58, No. 6, pp. 2974–82, 2010.

[100] P. Flandrin, N. Pustelnik, and P. Borgnat, "On Wigner-based sparse time-frequency distributions," in *Proceedings of IEEE International Workshop on Computational Advances in Multi-Sensor Adaptive Processing CAMSAP-15*, Cancun, 2015.

[101] T. Oberlin, S. Meignen, and V. Perrier, "Second-order synchrosqueezing transform or invertible reassignment? Towards ideal time-frequency representations," *IEEE Trans. Signal Proc.*, Vol. 63, No. 5, pp. 1335–44, 2015.

[102] D. H. Pham and S. Meignen, "High-order synchrosqueezing transform for multicomponent signals analysis – With an application to gravitational-wave signal," *IEEE Trans. Signal Proc.*, Vol. 65, No. 12, pp. 3168–78, 2017.

[103] N. E. Huang and S. P. Chen (eds.), *Hilbert-Huang Transform and Its Applications* (2nd edn.), Singapore: World Scientific, 2014.

[104] N. E. Huang, Z. Shen, S. R. Long, M. C. Wu, H. H. Shih, Q. Zheng, N.-C. Yen, C. C. Tung, and H. H. Liu, "The Empirical Mode Decomposition and the Hilbert Spectrum for nonlinear and nonstationary time series analysis," *Proc. Roy. Soc. A*, Vol. 454, pp. 903–95, 1998.

[105] I. Daubechies, J. Lu, and H.-T. Wu, "Synchrosqueezed wavelet transforms: An empirical mode decomposition-like tool," *Appl. Comp. Harm. Anal.*, Vol. 30, No. 1, pp. 243–61, 2011.

[106] G. Rilling, P. Flandrin, and P. Gonçalves, "On Empirical Mode Decomposition and its algorithms," in *Proceedings of IEEE-EURASIP Workshop on Nonlinear Signal and Image Processing NSIP-03*, Grado (Italy), 2003.

[107] M. Colominas, G. Schlotthauer, M. E. Torres, and P. Flandrin, "Noise-assisted EMD methods in action," *Adv. Adapt. Data Anal.*, Vol. 4, No. 4, pp. 1250025.1–1250025.11, 2013.

[108] É. Chassande-Mottin, I. Daubechies, F. Auger, and P. Flandrin, "Differential reassignment," *IEEE Signal Proc. Lett.*, Vol. 4, No. 10, pp. 293–4, 1997.

[109] V. Bargmann, "On a Hilbert space of analytic functions and an associated integral transform," *Comm. Pure Appl. Math.*, Vol. 14, pp. 187–214, 1961.

[110] F. Auger, É. Chassande-Mottin, and P. Flandrin, "On phase-magnitude relationships in the Short-Time Fourier Transform," *IEEE Signal Proc. Lett.*, Vol. 19, No. 5, pp. 267–70, 2012.

[111] F. Auger, É. Chassande-Mottin, and P. Flandrin, "Making reassignment adjustable: The Levenberg-Marquardt approach," in *Proceedings of IEEE International Conference on Acoustics, Speech and Signal Processing of ICASSP-12*, pp. 3889–92, Kyoto, 2012.

[112] M. Hansson-Sandsten and J. Brynolfsson, "The scaled reassigned spectrogram with perfect localization for estimation of Gaussian functions," *IEEE Signal Proc. Lett.*, Vol. 22, No. 1, pp. 100–4, 2015.

[113] Y. Lim, B. Shinn-Cunningham, and T. J. Gardner, "Sparse contour representation of sound," *IEEE Signal Proc. Lett.*, Vol. 19, No. 10, pp. 684–7, 2012.

[114] É. Chassande-Mottin, "Méthodes de réallocation dans le plan temps-fréquence pour l'analyse et le traitement de signaux non stationnaires," PhD thesis, Université de Cergy-Pontoise, France, 1998.

[115] T. J. Gardner and M. O. Magnasco, "Sparse time-frequency representations," *Proc. Natl. Acad. Sci.*, Vol. 103, No. 16, pp. 6094–9, 2006.

[116] S. Meignen, T. Gardner, and T. Oberlin, "Time-frequency ridge analysis based on reassignment vector," *Proceedings of 23rd European Signal Processing Conference EUSIPCO-15*, pp. 1486–90, Nice, 2015.

[117] S. Meignen, T. Oberlin, Ph. Depalle, P. Flandrin, and S. McLaughlin, "Adaptive multimode signal reconstruction from time-frequency representations," *Proc. Phil. Trans. Roy. Soc. A*, Vol. 374: 20150205, 2016.

[118] N. Delprat, B. Escudié, Ph. Guillemain, R. Kronland-Martinet, Ph. Tchamitchian, and Bruno Torrésani, "Asymptotic wavelet and Gabor analysis: Extraction of instantaneous frequencies," *IEEE Trans. Inf. Theory*, Vol. 38, No. 2, pp. 644–64, 1992.

[119] R. Carmona, W.-L. Hwang, and B. Torrésani, *Practical Time-Frequency Analysis: Gabor and Wavelet Transforms With an Implementation in S*, San Diego, CA: Academic Press, 1998.

[120] R. Bardenet, J. Flamant, and P. Chainais, "On the zeros of the spectrogram of white noise," arXiv:1708.00082v1, 2017.

[121] P. Flandrin, "On spectrogram local maxima," in *Proceedings IEEE International Conference on Acoustics, Speech and Signal Processing ICASSP-07*, pp. 3979–83, New Orleans, 2017.

[122] J. Conway and N. J. A. Sloane, *Sphere Packings, Lattices and Groups* (3rd edn.), New York: Springer-Verlag, 1999.

[123] P. J. Diggle, *Statistical Analysis of Spatial Point Patterns* (2nd edn.), London: Academic Press, 2003.

[124] D. Stoyan, W. S. Kendall, and J. Mecke, *Stochastic Geometry and Its Applications*, New York: John Wiley & Sons, 1995.

[125] L. S. Churchman, H. Flyvberg, and J. A. Spudich, "A non-Gaussian distribution quantifies distances measured with fluorescence localization techniques," *Biophys. J.*, Vol. 90, pp. 668–71, 2006.

[126] M. Abramowitz and I. Stegun, *Handbook of Mathematical Functions*, London: Dover, 1965.

[127] P. Embrechts, C. Klüppelberg, and T. Mikosch, *Modeling Extremal Events for Insurance and Finance*, Berlin-Heidelberg: Springer-Verlag, 1997.

[128] S. Coles, *An Introduction to Statistical Modeling of Extreme Values*, London: Springer, 2001.

[129] P. Calka, "An explicit expression for the distribution of the number of sides of the typical Poisson-Voronoi cell," *Adv. Appl. Prob.*, Vol. 35, pp. 863–70, 2003.

[130] J.-S. Ferenc and Z. Néda, "On the size distribution of Poisson Voronoi cells," *Physica A*, Vol. 385, pp. 518–26, 2007.

[131] M. Tanemura, "Statistical distributions of Poisson-Voronoi cells in two and three dimensions," *Forma*, Vol. 18, pp. 221–47, 2003.

[132] V. Lucarini, "From symmetry breaking to Poisson point process in 2D – Voronoi tessellations: The generic nature of hexagons," *J. Stat. Phys.*, Vol. 130, pp. 1047–62, 2008.

[133] A. Wang, "An industrial-strength audio search algorithm," in *Proceedings of Fourth International Conference on Music Information Retrieval ISMIR-03*, Baltimore, MD, 2003.

[134] S. Klimenko, I. Yakushin, A. Mercer, and G. Mitselmakher, "A coherent method for detection of gravitational wave bursts," *Class. Quant. Grav.*, Vol. 25, No. 11, pp. 114029, 2008.

[135] M. Haenggi, *Stochastic Geometry for Wireless Networks*, Cambridge: Cambridge University Press, 2013.

[136] R. P. Boas, *Entire Functions*, New York: Academic Press, 1954.

[137] M. Toda, "Phase retrieval problem in quantum chaos and its relation to the origin of irreversibility I," *Physica D*, Vol. 59, pp. 121–41, 1992.

[138] J. B. Hough, M. Krishnapur, Y. Peres, and B. Virág, *Zeros of Gaussian Analytic Functions and Determinantal Point Processes*, AMS University Lecture Series, Vol. 51, 2009.

[139] P. Lebœuf and A. Voros, "Chaos-revealing multiplicative representation of quantum eigenstates," *J. Phys. A: Math. Gen.*, Vol. 23, pp. 1765–74, 1990.

[140] H. J. Korsch, C. Müller, and H. Wiescher, "On the zeros of the Husimi distribution," *J. Phys. A: Math. Gen.*, Vol. 30, pp. L677–L684, 1997.

[141] O. Macchi, "The coincidence approach to stochastic point processes," *Adv. Appl. Prob.*, Vol 7, No. 2, pp. 83–122, 1975.

[142] C. A. N. Biscio and F. Lavancier, "Quantifying repulsiveness of determinantal point processes," *Bernoulli*, Vol. 22, No. 4, pp 2001–28, 2016.

[143] F. Lavancier, J. Møller, and E. Rubak, "Determinantal point process models and statistical inference," *J. Roy. Stat. Soc. B*, Vol. 77, No. 4, pp. 853–77, 2015.

[144] J. H. Hannay, "Chaotic analytic zero points: Exact statistics for those of a random spin state," *J. Phys. A: Math. Gen.*, Vol. 29, pp. L101–L105, 1996.

[145] F. Nazarov, M. Sodin, and A. Volberg, "Transportation to random zeroes by the gradient flow," *Geom. Funct. Anal.*, Vol. 17, No. 3, pp. 887–935, 2007.

[146] P. Flandrin, "Time-frequency filtering from spectrogram zeros," *IEEE Signal Proc. Lett.*, Vol. 22, No. 11, pp. 2137–41, 2015.

[147] P. Balasz, D. Bayer, F. Jaillet, and P. Søndergaard, "The pole behavior of the phase derivative of the short-time Fourier transform," *Appl. Comp. Harm. Anal.*, Vol. 40, pp. 610–21, 2016.

[148] J. F. Nye and M. V. Berry, "Dislocations in wave trains," *Proc. Roy. Soc. Lond. A*, Vol. 336, pp. 165–90, 1974.

[149] L. Blanchet, T. Damour, B. R. Iyer, C. M. Will, and A. G. Wiseman, "Gravitational-radiation damping of compact binary systems to second post-Newtonian order," *Phys. Rev. Lett.*, Vol. 74, No. 18, pp. 3515–18, 1995.

[150] B. S. Sathyaprakash and S. V. Dhurandhar, "Choice of filters for the detection of gravitational waves from coalescing binaries," *Phys. Rev. D*, Vol. 44, pp. 3819–34, 1991.

[151] K. S. Thorne, "Gravitational radiation," in *300 Years of Gravitation* (S. W. Hawking and W. Israel, eds.), pp. 330–458, Cambridge: Cambridge University Press, 1987.

[152] A. Buonanno and T. Damour, "Effective one-body approach to general relativistic two-body dynamics," *Phys. Rev. D*, Vol. 59, pp. 084006, 1999.

[153] É. Chassande-Mottin and P. Flandrin, "On the time-frequency detection of chirps," *Appl. Comp. Harm. Anal.*, Vol. 6, No. 2, pp. 252–81, 1999.

[154] S. Jaffard, J. L. Journé, and I. Daubechies, "A simple Wilson orthonormal basis with exponential decay," *SIAM J. Math. Anal.*, Vol. 22, pp. 554–73, 1991.

[155] B. P. Abbott et al., "GW151226: Observation of gravitational waves from a 22-solar-mass binary black hole coalescence," *Phys. Rev. Lett.*, Vol. 116, pp. 241103-1–241103-14, 2016.

[156] P. Flandrin, "The sound of silence: Recovering signals from time-frequency zeros," in *Proceedings of Asilomar Conference on Signals, Systems, and Computers*, pp. 544–8, Pacific Grove, CA, 2016.

[157] É. Chassande-Mottin and P. Flandrin, "On the time-frequency detection of chirps and its application to gravitational waves," in *Second Workshop on Gravitational Wave Data Analysis*, pp. 47–52, Paris: Éditions Frontières, 1999.

[158] P. E. Nachtigall and P. W. B. Moore (eds.), *Animal Sonar: Processes and Performance*, NATO ASI Series A: Life Sciences, Vol. 156, New York: Plenum Press, 1988.

[159] T. Nagel, "What is it like to be a bat?," *Phil. Rev.*, Vol. 83, No. 4, pp. 435–50, 1974.

[160] R. A. Altes and E. L. Titlebaum, "Bat signals as optimally Doppler tolerant waveforms," *J. Acoust. Soc. Amer.*, Vol. 48(4B), pp. 1014–20, 1970.

[161] J. A. Simmons, "The resolution of target range by echolocating bats," *J. Acoust. Soc. Amer.*, Vol. 54, pp. 157–73, 1973.

[162] J. A. Simmons, "Perception of echo phase information in bat sonar," *Science*, Vol. 207, pp. 1336–8, 1979.

[163] J. A. Simmons, M. Ferragamo, C. F. Moss, S. B. Stevenson, and R. A. Altes, "Discrimination of jittered sonar echoes by the echolocating bat, *Eptesicus fuscus*: The shape of targets images in echolocation," *J. Comp. Physiol. A*, Vol. 167, pp. 589–616, 1990.

[164] B. Møhl, "Detection by a pipistrelle bat of normal and reverse replica of its sonar pulses," *Acustica*, Vol. 61, pp. 75–82, 1986.

[165] R. A. Altes, "Some theoretical concepts for echolocation," in [158], pp. 725–52.

[166] R. A. Altes, "Detection, estimation, and classification with spectrograms," *J. Acoust. Soc. Amer.*, Vol. 67, No. 4, pp. 1232–46, 1980.

[167] R. A. Altes, "Echolocation as seen from the viewpoint of radar/sonar theory," in *Localization and Orientation in Biology and Engineering* (D. Varjú and H.-U. Schnitzler, eds.), pp. 234–44, Berlin: Springer-Verlag, 1984.

[168] P. Saillant, J. A. Simmons, and S. Dear, "A computational model of echo processing and acoustic imaging in frequency-modulated echolocating bats: The spectrogram correlation and transformation receiver," *J. Acoust. Soc. Amer.*, Vol. 94, pp. 2691–712, 1993.

[169] S. Jaffard and Y. Meyer, "Wavelet methods for pointwise regularity and local oscillations of functions, *Memoirs Amer. Math. Soc.*, Vol. 123, No. 587, 1996.

[170] K. Falconer, *Fractal Geometry*, New York: John Wiley & Sons, 1990.

[171] M. V. Berry and Z. V. Lewis, "On the Weierstrass-Mandelbrot fractal function," *Proc. Roy. Soc. London A*, Vol. 370, pp. 459–84, 1980.

[172] P. Borgnat and P. Flandrin, "On the chirp decomposition of Weierstrass-Mandelbrot functions, and their time-frequency interpretation," *Appl. Comp. Harm. Anal.*, Vol. 15, pp. 134–46, 2003.

[173] J. Bertrand, P. Bertand, and J.-Ph. Ovarlez, "The Mellin transform," in *The Transforms and Applications Handbook* (A. Poularikas, ed.), Boca Raton, FL: CRC Press, 1990.

[174] J. Flamant, N. Le Bihan, and P. Chainais, "Time-frequency analysis of bivariate signals," arXiv:1609.02463, 2016.

[175] B. Roussel, C. Cabart, G. Fève, E. Thibierge, and P. Degiovanni, "Electron quantum optics as quantum signal processing," *Phys. Status Solidi B*, Vol. 254, pp. 16000621, 2017.

[176] L. J. Stanković, M. Daković, and E. Sejdić, "Vertex-frequency analysis: A way to localize graph spectral components," *IEEE Signal Proc. Mag.*, Vol. 34, No. 4, pp. 176–82, 2017.

Index

Altes, R. A., 182, 188
ambiguity function, 41–7, 50, 51, 92–5, 181
Amplitude Modulation, 19, 48, 63, 112
analytic
    analytic noise, 28, 120
    analytic signal, 18, 27
    Gaussian Analytic Functions, 6, 143, 145,
        146, 151
attractor, 111, 112, 117

Bargmann, V., 106, 139, 142
    Bargmann factorization, 139
    Bargmann transform, 33, 107, 140–2, 162, 163
basin, 82, 111–3, 115, 134, 147, 148, 160, 162
bat, 11–3, 19, 20, 175–88, 192
Beckner, W., 38
Berry, M. V., 14
Blanc-Lapierre, A., 2
Boltzmann, L., 38

chirp, 9–11, 14–21, 38, 48, 54–6, 61–3, 75,
        77–82, 86–8, 90, 91, 94–7, 100, 101, 103, 104,
        111, 112, 114, 115, 139, 154, 155,
        157, 168, 169, 171–7, 179–88, 191, 192,
        194–6
circular, 27, 28, 42, 106, 120, 122, 123, 142
Claasen, T. A. C. M., 49
Cohen, L., 46, 92
    Cohen's class, 46, 47, 53, 83, 92
contour, 114, 115, 160, 162, 167
correlation, 6, 40, 42, 174, 184
    autocorrelation, 92, 183, 185
    correlation function, 26, 27, 186
    correlation interpretation, 26, 44, 45, 92
    correlation radius, 42
    correlation structure, 118, 155, 181
    cross-correlation, 92, 172, 181, 182
    deterministic correlation, 25
    microscopic correlation, 26, 84
    pair correlation function, 144–6
    spectrogram correlation, 188
    stationary correlation function, 45
    time-frequency correlation, 42–4, 92, 121

data
    big data, 5
    data analysis, 1, 3
    data-driven, 66, 98, 99, 105, 114
    data science, 1, 4
    small data, 9
    surrogate data, 61–6, 85
Daubechies, I., 4, 89, 99
Delaunay, B., 149
    Delaunay triangles, 149–52, 154–6, 158
    Delaunay triangulation, 149, 150, 153, 155,
        157–60, 172, 173
denoising, 11, 21, 153
determinantal point process, 145
Dirac, P. A. M., 24, 190
    Dirac's δ-function, 24, 190
Doppler
    Doppler effect, 20, 44
    Doppler tolerance, 183–5
Dunn, H. K., xiii

echolocation, 12–4, 20, 175, 176, 178–80, 183,
        186–8
Einstein, A., 9
Empirical Mode Decomposition, 83, 98, 99, 101,
        103, 105
entropy, 38, 51, 54
    Rényi entropies, 38
    Shannon entropy, 38

factorization, 139
    Bargmann factorization, 106, 139, 163
    Weierstrass-Hadamard factorization, 140,
        142, 163
Fortet, R., 2
Fourier, J., 3–6, 20, 24, 29, 31, 32, 35, 39, 44, 45,
        48–50, 56, 81, 82, 92–4, 107, 184
    Fast Fourier Transform, 3, 4, 29
    Fourier analysis, 3, 4, 39, 44
    Fourier integral, 3
    Fourier modes, 31, 61, 62, 105
    Fourier series, 3, 29, 192, 193
    Fourier spectrum, 32, 61

Fourier transform, 5, 22, 23, 25, 26, 30, 31, 33, 35, 36, 41, 44–7, 61, 77, 84, 92–4, 118, 156, 184, 194
  self-Fourier, 31
  Short-Time Fourier Transform, 33, 40
frequency, 4, 5, 16–8, 29, 32, 35
  Frequency Modulation, 13, 63, 111, 112
  instantaneous frequency, 9, 16–9, 26, 37, 49, 55, 77, 78, 83, 87–90, 97, 102, 104, 112, 114, 154, 168–70, 173, 174, 176, 179, 180, 182, 184, 185, 192, 194

Gabor, D., 32, 33, 35, 37
  Gabor analysis, 4
  Gabor expansion, 4, 37, 171
  Gabor logon, 37, 69, 141
Gauss, C. F., 29, 31, 32
  colored Gaussian noise, 158, 160
  complex white Gaussian noise, 119, 125, 131, 133, 146
  Gaussian bump, 121
  Gaussian distribution, 126, 156
  Gaussian spectrogram, 48, 69, 70, 72, 73, 75, 80, 81, 91, 93, 95, 102, 104, 117
  Gaussian waveform, 30, 36, 43, 48, 86
  Gaussian window, 158, 195
  white Gaussian noise, 24, 65, 86, 101, 117, 118, 122, 126, 129, 131, 134–6, 141, 148, 149, 152, 153, 155, 156, 158
Gauss, C. F., 29
  analytic white Gaussian noise, 28
  circular Gaussian window, 42, 54, 92, 106–8, 113, 114, 119–21, 142, 163, 172
  colored Gaussian noise, 156, 159
  complex white Gaussian noise, 28, 116–21, 124, 129–32, 135, 136, 139, 142–5, 152, 153, 156
  complex-valued white Gaussian noise, 26, 27
  fractional Gaussian noise, 157
  Gaussian Analytic Functions, 6, 143, 145, 146, 151
  Gaussian bump, 122
  Gaussian case, 6, 25, 27, 41, 42, 53, 84
  Gaussian distribution, 29, 124, 135, 142, 145, 152
  Gaussian envelope, 38, 61
  Gaussian expansion, 42
  Gaussian fluctuation, 127
  Gaussian kernel, 121
  Gaussian matrices, 145
  Gaussian process, 56
  Gaussian spectrogram, 34, 53, 69, 74, 85, 87, 102, 116, 120, 141
  Gaussian waveform, 6, 29–33, 35, 37–40, 42, 45, 51–3, 69, 70, 77
  Gaussian window, 34, 42, 52, 53, 69, 85, 140, 156
  Gaussianity, 26, 116, 129
  real-valued white Gaussian noise, 24, 120

white Gaussian noise, 24, 26, 65, 74, 84–6, 101, 116, 120, 142, 154, 157, 171, 181
Gendrin, R., 83
Ginibre ensemble, 145, 146
golden triangle, 2–5, 23, 48, 49, 105
Gröchenig, K., 50
gravitational wave, 9–11, 14, 16, 20, 138, 168–73, 184, 187, 192
Griffin, D.W., 13, 175
Grossmann, A., 4
Gumbel, E. J., 132
  Gumbel distribution, 132–4

Hannay, J. H., 145, 146
Heisenberg, W., 35, 39, 81
  Heisenberg uncertainty principle, 35
Hermite, C., 26
  Hermite function, 31, 52, 85–8, 90, 91, 94, 95, 141, 142, 160, 162
  Hermite polynomials, 141
  Hermite reassigned spectrogram, 88
  Hermite spectrogram, 86, 87
  Hermite window, 87, 88
Hilbert, D., 6, 102–4
  Hilbert transform, 18, 27, 100
  Hilbert-Huang Transform, 6, 98, 102, 104, 105
Hirschman Jr, I. I., 38
Huang, N. E., 99, 100, 105
  Hilbert-Huang Transform, 6, 98, 100, 102, 104, 105
  Huang's algorithm, 100
Hudson, R. L., 45
Husimi distribution, 34, 69, 143

interference, 61
  interference geometry, 55, 71, 74, 188
  interference pattern, 9, 90, 91, 188
  interference structure, 71, 81
  interference terms, 70–4, 77–82, 92, 93
Intrinsic Mode Function, 99

Jaffard, S., 192

Kodera, K., 83
Koenig, W., xiii

Lacy, D. Y., xiii
lattice, 6, 122–30, 134, 136, 137, 150, 152
Lieb, E., 51
LIGO interferometer, 9, 10, 171, 172
localization, 39, 54, 55, 77–9, 81, 87, 88, 90, 94, 96, 97, 179, 187, 188

Macchi, O., x, 145
Maes, S., 89
Mallat, S., 4
Mandelbrot, B. B., 57, 193, 194

matched filter, 42, 171–4, 176, 181, 182, 184–6
Mecklenbräuker, W. F. G., 49
Mellin, H., 194
　Mellin chirp, 194
　Mellin transform, 194
Meyer, Y., x, 4, 192
　Meyer wavelet, 171
Morlet, J., 4, 20
　Morlet wavelet, 58
Moyal's formula, 46, 50, 186
multitaper, 60, 65, 83, 85–8
Møhl, B., 186

Newton, I., 49
　Newton algorithm, 109
　Newtonian approximation, 168
Nyquist, H., 31, 32

Parseval's relation, 22, 35, 50
Penzias, A., 1
phase
　phase derivative, 18, 37, 109, 162–4
　phase diagram, 165–7
　phase dislocation, 164–7
　phase singularities, 165, 166
　phase spectrum, 61–3
　phase transition, 137
Pierce, G. W., 13, 175
proper process, 27

Rényi, A., 38
　Rényi entropies, 38
reassignment, 6, 49, 78–84, 88–91, 96–8, 107–9,
　　111, 112, 114, 115, 117, 146–8, 162
　adjustable reassignment, 109, 110
　differential reassignment, 108, 146
　multitaper reassignment, 83, 88
　reassigned spectrogram, 79, 80, 82
　reassignment operators, 83, 89, 108, 109
　reassignment principle, 83
　reassignment vector field, 6, 82, 97, 108, 111–5,
　　117, 162
relation function, 27, 120
repeller, 111, 113, 117, 162
reproducing kernel, 41–3, 47, 71, 73, 107, 120, 121,
　　123, 124, 138, 142, 145, 155
ridge, 109, 114, 115
Riemann, B., 14, 188, 192
　Riemann hypothesis, 15, 190
　Riemann zeros, 190
　Riemann's psi function, 189
　Riemann's sigma function, 192, 193
　Riemann's zeta function, 14, 188
　Riemann-Siegel expansion, 188, 189

scalogram, 47, 85
Scott de Martinville, É. -L., xiii

Shannon, C. E., 31, 32, 38
　Shannon entropy, 38
signal
　analytic signal, 18, 27
　signal processing, 1–4, 9, 21, 29, 33, 38, 49, 56,
　　57, 83, 105
Simmons, J. A., 184, 186, 188
singularities, 162, 164, 190, 192
Spallanzani, L., 13
sparsity, 6, 90, 92–6, 171
spectrogram, 6, 40, 46–8, 52, 53, 59, 60, 62, 69, 70,
　　77, 78, 84, 106
squeezed state, 38, 54
stationarity, 6, 25, 45, 56, 57, 61, 62, 118, 120–2,
　　135, 143
　local stationarity, 56
　quasi-stationarity, 56
　relative stationarity, 57, 58, 65
　second-order stationarity, 25, 56, 119, 120
　stationarity test, 56, 57, 60, 64, 65
surrogate data, 61–6, 85, 172
synchrosqueezing, 6, 83, 88–91, 96–9, 105, 159

Thomson, D. J., 85
Titlebaum, E. L., 182

uncertainty, 35–7, 39, 50, 54, 55, 80, 92, 107, 184
　entropic uncertainty, 38
　Heisenberg uncertainty principle, 35
　time-frequency uncertainty, 35, 37, 50, 53, 54, 79
　uncertainty interpretation, 39, 42
　uncertainty relation, 6, 40, 49, 54
　variance-based uncertainty, 38, 52

Ville, J. A., 45, 49
　Wigner-Ville distribution, 44
Villedary, C. de, 83
Virgo, 171
Voronoi, G., 73, 134, 145
　Voronoi cell, 74, 75, 135–40, 146–50, 152
　Voronoi tessellation, 74, 134–6, 139, 140, 145,
　　147, 148, 150, 151

wavelet, 4, 5, 20, 47–9, 57–9, 83, 85, 89, 99, 105,
　　171, 172, 192
Weierstrass, K., 192
　Weierstrass function, 16, 191, 192, 195
　Weierstrass-Hadamard factorization, 140, 163
　Weierstrass-Mandelbrot function, 194, 195
Welch, P. D., 84
　Welch procedure, 60, 84, 85
Wiener, N., 2, 4
　Wiener-Khintchine-Bochner theorem, 25,
　　44, 118
Wigner, E. P., 43
　Wigner distribution, 48, 70, 72, 73, 80–2, 91,
　　93, 95

Wigner interference terms, 81
Wigner-Ville distribution, 44
Wigner, E. P., 49
  smoothed Wigner distribution, 53, 106, 187
  Wigner distribution, 44–54, 59, 60, 70, 71, 77–80, 83, 90, 92, 93, 105, 187
  Wigner spectrum, 45, 59
Wilson, R. W., 1
window, 47, 60, 84–6, 92, 106, 108, 118, 186, 191, 196
  circular Gaussian window, 42, 54, 92, 106–8, 113, 114, 119–21, 142, 156, 158, 163, 172
  Gabor window, 4
  Gaussian window, 34, 42, 52, 53, 69, 85, 140, 195
  Hermite window, 85, 87, 88
  modified moving window, 83

multitaper windows, 85
observation window, 169
self-window, 47
short-time window, 33, 40, 41, 43, 46, 64, 65, 77, 79, 86, 88, 171
time-frequency window, 112, 196

zeros
  Chaotic Analytic Zeros Points, 143
  Riemann zeros, 190
  spectrogram zeros, 6, 15, 70–5, 112–5, 117, 118, 138–63, 165, 172, 173, 188
  STFT zeros, 162, 164–6
  zeros-based algorithm, 6
  zeros-based filtering, 113, 155, 158–61
  zeros-based reconstruction, 161
  zeta function zeros, 14, 15, 189, 190